R实战

系统发育树的
数据集成操作及可视化

Data Integration,
Manipulation and Visualization of Phylogenetic Trees

余光创　著
李林　罗晓　译

电子工业出版社
Publishing House of Electronics Industry
北京·BEIJING

内 容 简 介

本书系统地介绍使用 treeio、tidytree、ggtree 和 ggtreeExtra 等 R 软件包操作系统发育树的全套流程，包括对树文件的解析，以及树与其相关数据的操作、整合、可视化等内容。

本书由余光创撰写，旨在为系统发育树的操作与呈现提供指导。如果读者需要进行系统发育树的相关操作，却又觉得无从下手，那么这本书会提供很大的帮助。关于系统发育树的大部分问题，都能在本书中找到答案。

未经许可，不得以任何方式复制或抄袭本书之部分或全部内容。
版权所有，侵权必究。

图书在版编目（CIP）数据

R 实战：系统发育树的数据集成操作及可视化／余光创著；李林，罗晓译．—北京：电子工业出版社，2023.4

ISBN 978-7-121-45182-9

Ⅰ．①R… Ⅱ．①余… ②李… ③罗… Ⅲ．①程序语言－程序设计 Ⅳ．①TP312

中国国家版本馆 CIP 数据核字（2023）第 041769 号

责任编辑：张慧敏　　　　　　　特约编辑：田学清
印　　　刷：中国电影出版社印刷厂
装　　　订：中国电影出版社印刷厂
出版发行：电子工业出版社
　　　　　北京市海淀区万寿路 173 信箱　　邮编：100036
开　　本：720×1000　　1/16　　印张：17.5　　字数：314 千字
版　　次：2023 年 4 月第 1 版
印　　次：2023 年 4 月第 1 次印刷
定　　价：109.00 元

凡所购买电子工业出版社图书有缺损问题，请向购买书店调换。若书店售缺，请与本社发行部联系，联系及邮购电话：（010）88254888，88258888。

质量投诉请发邮件至 zlts@phei.com.cn，盗版侵权举报请发邮件至 dbqq@phei.com.cn。
本书咨询联系方式：（010）51260888-819，faq@phei.com.cn。

关于作者

余光创，生物信息学教授，在香港大学公共卫生学院获得博士学位，现任南方医科大学生物信息学系系主任。作为一位活跃的 R 语言用户，他编写了许多 R 软件包，如 aplot、badger、ChIPseeker、clusterProfiler、DOSE、emojifont、enrichplot、ggbreak、ggfun、ggimage、ggplotify、ggtree、GOSemSim、hexSticker、meme、meshes、nCov2019、plotbb、ReactomePA、scatterpie、seqmagick、seqcombo、shadowtext、tidytree 及 treeio，同时指导学生开发了一系列 R 软件包，如 ggmsa、ggtreeExtra、MicrobiomeProfiler 及 MicrobiotaProcess 等。

他的课题组旨在通过开发新的软件工具和对生物医学数据的新分析，对人类健康及疾病产生新的见解。他的课题组开发的软件包能够帮助生物学家分析数据，并揭示隐藏在数据之中的生物学线索。

余光创发表了多篇期刊论文，包括 5 篇高被引论文[1-5]，被引用超过了 10000 次。其关于 ggtree 的论文[1]被选为专题文章①，以庆祝 *Methods in Ecology and Evolution* 期刊创刊 10 周年。他连续两年（2020 年和 2021 年）入选生物医学工程领域中国高被引学者。

参考文献

[1] Yu G, Smith D K, Zhu H, et al. ggtree: an R package for visualization and annotation of phylogenetic trees with their covariates and other associated data[J]. Methods in Ecology and Evolution, 2016, 8(1): 28-36.

[2] Yu G, Wang L, Han Y, et al. clusterProfiler: an R package for comparing biological themes among gene clusters[J]. OMICS: A Journal of Integrative Biology, 2012, 16(5): 284-287.

[3] Yu G, Wang L, He Q. ChIPseeker: an R/Bioconductor package for ChIP peak annotation, comparison and visualization[J]. Bioinformatics, 2015, 31(14): 2382-2383.

[4] Yu G, Wang L, Yan G, et al. DOSE: an R/Bioconductor package for disease ontology semantic and enrichment analysis[J]. Bioinformatics, 2015, 31(4): 608-609.

[5] Yu G H Q Y. ReactomePA: an R/Bioconductor package for reactome pathway analysis and visualization. [J]. Molecular BioSystems, 2016, 12(2): 477-479.

① 10 周年第八卷：多变量数据与系统发育树可视化 / 详情参见"外链资源"文档中关于作者第 1 条

推荐序

进化论的提出是科学史上的一次革命，就像杜布赞斯基说的"如果没有进化之光，生物学的一切都将无法理解"。有趣的是，作为进化关系最直观表述的系统发育树却早于进化论的诞生而被创造出来。人们通过整理达尔文留下的笔记发现，早在1837年达尔文还在为撰写《物种起源》整理旅行日记时，就在笔记本上绘制了第一张"生命之树"，用于描述物种的演变过程。因此可以说，系统发育树与进化论同根同源，既是进化论思想的呈现，又是创造进化论的思维工具。

达尔文的灵光乍现为生命科学的研究提供了一个强大的工具。今天，系统发育树的计算生成和可视化已经成为生命科学研究中不可或缺的分析技术。本书的作者余光创很早就致力于生物信息分析技术的研究，十余年间发布了近30个具有影响力的分析软件，其中很多已经成为生命科学研究中的经典分析工具。在系统发育树分析方面，由余光创团队开发的ggtree软件包获得了巨大成功，成为进化生物学、生态学等领域十分重要的分析工具。

本书详细地介绍了使用R语言及相关软件包进行系统发育树分析及可视化的方法。我相信，与书中介绍的各种知名软件包一样，这本书也将成为从事生命科学相关研究的科学家、研究生和生物信息工程师的必备工具，一定能帮助读者更好地绘制属于自己的"生命之树"。

伯晓晨
军事科学院军事医学研究院

前言

撰写这本书是为了给使用 tidytree、treeio、ggtree 与 ggtreeExtra 这一套 R 包进行系统发育树数据整合及可视化等操作的用户提供一个指南。因此，我们假定，阅读本书的读者具有一定的 R 语言及 ggplot2 包的使用基础。

ggtree 包的开发始于我在香港大学攻读博士期间。在那时，我加入了新发传染性疾病国家重点实验室（State Key Laboratory of Emerging Infectious Diseases，SKLEID），并在管轶教授与林赞育副教授的指导下，参与了修改 Newick 树字符串的工作，使其能在系统发育树的内部节点标签中包含一些额外的信息，如氨基酸替换，以进行可视化。我编写了一个 R 脚本来实现这个功能，但很快意识到大多数的系统发育树可视化软件只能通过节点标签展示单一类别的数据。在那时，我们基本不可能同时做到展示两个数据变量来进行比较分析。人们往往需要借助图像后期处理软件才能绘制出能同时展示不同分支或节点相关数据（比如自举值或替换信息）的树图，这使得我萌生了开发 ggtree 包的念头。首先，我认为一个好的用户界面必须要完全支持 ggplot2 包通过叠加图层来绘图的语法。这样，简单的图绘制很容易，而复杂的图只不过是简单图的组合。

经过数年的开发，ggtree 已经进化为一个软件包套组，其中包括通过整洁接口（tidy interface）来操作树及相关数据的 tidytree，用于输入及输出含有丰富注释数据的树文件的 treeio，用于可视化及注释树的 ggtree，以及用于将数据展示在矩形布局树的右侧或环形布局树的外圈的 ggtreeExtra。ggtree 是一个通用工具，支持多种不同种类的树与树形结构，同时能被应用于多个不同的学科，帮助科研工作者在进化结构或层次结构的背景下呈现并解读数据。

本书的结构

- 第 1 篇　树数据的输入 / 输出及操作。

第 1 ～ 3 章：主要介绍用于树数据输入 / 输出的 treeio 包，以及用于树数据操作的 tidytree 包。

- 第 2 篇　树数据的可视化及注释。

第 4 ～ 8 章：主要介绍如何使用在 ggtree 包中实现的图形语法进行树的可视化与注释，以及如何将树的关联数据呈现在树上。

- 第 3 篇　ggtree 拓展包。

第 9 ～ 11 章：主要介绍用于在环形布局的树上呈现数据的 ggtreeExtra 包，以及其他的一些拓展包，如 MicrobiotaProcess 和 tanggle 等。

- 第 4 篇　杂项。

第 12 ～ 13 章：主要介绍一些由 ggtree 套组提供的实用工具，以及一系列可复现的示例。

软件的信息及约定

编译本书时的 R 语言及核心包的信息如下。

```
R.version.string
```

```
## [1] "R version 4.1.2 (2021-11-01)"
```

```
library(treedataverse)
```

```
##   Attaching packages    treedataverse 0.0.1
```

```
## ape          5.5       treeio       1.18.1
## dplyr        1.0.7     ggtree       3.2.1
## ggplot2      3.3.5     ggtreeExtra  1.4.1
## tidytree     0.3.6
```

其中，treedataverse 是一个元包（meta package），能帮助我们轻松地安装并加载本书中所介绍的用于树的处理及可视化的核心包。我们可以在附录 A 中找到安装 treedataverse 的指南。

本书的数据集有以下 3 个来源。

（1）模拟数据。

（2）R 包中的数据集。

（3）从互联网上下载的数据。

为了使用户更容易地获取互联网中的数据，我们将这些数据集与能获取到其来源的详细信息存储到 TDbook 包中，这些信息包括 URL、作者信息及引文信息。TDbook 包已经被发布在 CRAN 上，而用户可以使用 install.packages("TDbook") 命令来安装 TDbook 包。

在本书中，函数名的后面都添加了一对括号（如 treeio::read.beast()），其中双冒号操作符 "::" 表示访问一个包中的对象。

致谢

在此，非常感谢徐双斌对 ggtree 系列软件包的长期维护，同时感谢李林与罗晓对本书英文版书稿的审译和校对。

我在开发 ggtree 软件套组期间获得了许多人的帮助。在此感谢 Hadley Wickham，因为他创建了 ggtree 包所依赖的 ggplot2 包；感谢管轶教授与林讚育副教授，在我攻读博士期间为 ggtree 包的开发提供了很多好的建议；感谢 Richard Ree 邀请我参加系统发育树可视化的催化会议；感谢 William Pearson

邀请我在 *Current Protocols in Bioinformatics* 期刊上发表一篇关于 ggtree 包的 protocol 文章；感谢徐双斌、夏永和、黄瑞珠、Justin Silverman、Bradley Jones、Watal M. Iwasaki、Casey Dunn、Tyler Bradley、Konstantinos Geles、Zebulun Arendsee 与其他许多对源代码做出过贡献或给予过我反馈意见的人；最后，我还想感谢所有 ggtree 邮件列表[①]组的成员，由于他们提出了很多具有挑战性的问题，所以才能改进 ggtree 软件套组的功能。

余光创

---

① ggtree 邮件列表详情请参见"外链资源"文档中前言第 1 条

目录

第 1 篇　树数据的输入 / 输出及操作

第 1 章　导入带有数据的树文件 ... 2
- 1.1　系统发育树构建概述 ... 2
- 1.2　系统发育树文件格式 ... 4
  - 1.2.1　Newick 树文件 ... 4
  - 1.2.2　NEXUS 格式 ... 5
  - 1.2.3　NHX 格式 ... 7
  - 1.2.4　Jplace 格式 ... 7
  - 1.2.5　利用软件输出文件 ... 8
- 1.3　使用 treeio 导入树及相关数据 ... 13
  - 1.3.1　treeio 简介 ... 17
  - 1.3.2　treeio 解析函数演示 ... 18
  - 1.3.3　将其他树形对象转换为 phylo 对象或 treedata 对象 ... 29
  - 1.3.4　从 treedata 对象中获取信息 ... 31
- 1.4　总结 ... 34
- 1.5　本章练习题 ... 35
- 参考文献 ... 35

第 2 章　操作含有关联数据的树 ... 38
- 2.1　使用 tidy 接口操作树数据 ... 38
  - 2.1.1　phylo 对象 ... 38

2.1.2　treedata 对象 40
　　2.1.3　访问相关节点 41
2.2　数据整合 43
　　2.2.1　整合树数据 43
　　2.2.2　将外部数据关联到系统发育树 46
　　2.2.3　对分类单元进行分组 48
2.3　重新设定树的根节点 51
2.4　重新调整分支标尺 55
2.5　对包含数据的树取子集 56
　　2.5.1　删除系统发育树中的叶节点 56
　　2.5.2　通过叶节点标签对树取子集 58
　　2.5.3　通过内部节点编号对树取子集 60
2.6　操作树数据以进行可视化 62
2.7　总结 65
2.8　本章练习题 65
参考文献 65

第3章　导出含有数据的树 67
3.1　简介 67
3.2　将树数据导出为 BEAST Nexus 格式的文件 68
　　3.2.1　软件输出文件的导出与转换 68
　　3.2.2　将树与外部数据结合 71
　　3.2.3　合并不同来源的树数据 72
3.3　将树数据导出为 jtree 格式的文件 74
3.4　总结 77
3.5　本章练习题 77
参考文献 77

第 2 篇　树数据的可视化及注释

第4章　系统发育树可视化 80
4.1　简介 80

4.2 使用 ggtree 包对系统发育树进行可视化 81
4.2.1 基本的系统发育树的可视化 82
4.2.2 系统发育树的布局 83
4.3 绘制树的构成部分 89
4.3.1 绘制树的标尺 89
4.3.2 绘制内/外部节点 91
4.3.3 绘制标签 91
4.3.4 绘制根分支 93
4.3.5 给树着色 94
4.3.6 调整进化树标尺 98
4.3.7 修改主题组件 100
4.4 对树列表进行可视化 100
4.4.1 使用不同变量的值注释同一棵树 102
4.4.2 密度树 103
4.5 总结 104
4.6 本章练习题 105
参考文献 105

第 5 章 系统发育树注释 107
5.1 使用图形语法对树进行可视化及注释 107
5.2 进化树注释图层 109
5.2.1 彩色条带 109
5.2.2 突出显示进化枝 112
5.2.3 连接分类单元 114
5.2.4 进化推论的不确定性 116
5.3 使用进化软件输出结果注释树 117
5.4 总结 120
5.5 本章练习题 121
参考文献 121

第 6 章 系统发育树的可视化探索 122
6.1 查看选定的进化枝 122

6.2	缩小选定的进化枝	124
6.3	折叠及展开进化枝	124
6.4	对分类单元进行分组	127
6.5	对系统发育树结构的探索	128
6.6	总结	133
6.7	本章练习题	133
参考文献		133

第 7 章 绘制含有数据的树 ... 134

7.1	将外部数据映射到树结构	134
7.2	基于树的结构将图与树对齐	136
7.3	对含有关联矩阵的树进行可视化	138
7.4	对含有多序列比对结果的树进行可视化	142
7.5	复合图	143
7.6	总结	145
7.7	本章练习题	147
参考文献		147

第 8 章 使用轮廓图和子图注释进化树 ... 148

8.1	使用图像注释进化树	148
8.2	使用 phylopic 注释进化树	149
8.3	使用子图注释进化树	150
	8.3.1 使用柱状图进行注释	151
	8.3.2 使用饼图进行注释	152
	8.3.3 使用多种不同类型的图表进行注释	152
8.4	玩转 phylomoji	153
	8.4.1 在环形布局或扇形布局的树中使用表情符号	155
	8.4.2 使用表情符号作为进化枝标签	156
	8.4.3 Apple 彩色表情符号	157
	8.4.4 使用 ASCII Art 呈现 phylomoji	158
8.5	总结	159

8.6 本章练习题 159

参考文献 159

第 3 篇　ggtree 拓展包

第 9 章　对其他树形对象使用 ggtree 包 162

9.1 使用 ggtree 包绘制系统发育树对象 162

9.1.1 phylo4 对象和 phylo4d 对象 162

9.1.2 phylog 对象 165

9.1.3 phyloseq 对象 166

9.2 使用 ggtree 包绘制树状图 169

9.3 使用 ggtree 包绘制树形网络图 171

9.4 使用 ggtree 包绘制其他树形结构 172

9.5 总结 173

9.6 本章练习题 174

参考文献 174

第 10 章　使用 ggtreeExtra 包在环形布局上呈现数据 175

10.1 简介 175

10.2 基于树的结构将图与树对齐 175

10.3 在多维数据的可视化中将多个图与树对齐 178

10.4 群体遗传学示例 183

10.5 总结 190

10.6 本章练习题 190

参考文献 191

第 11 章　其他 ggtree 扩展包 192

11.1 使用 MicrobiotaProcess 包进行分类学注释 193

11.2 使用 tanggle 包可视化系统发育网络图 194

11.3 总结 195

11.4 本章练习题 196

参考文献 196

第4篇 杂项

第12章 ggtree 包中的实用工具 198
12.1 分面相关实用工具 198
12.1.1 facet_widths() 函数 198
12.1.2 facet_labeller() 函数 200
12.2 几何对象图层 201
12.3 布局相关工具 202
12.4 标尺相关工具 203
12.4.1 扩大指定面板的 x 轴范围 203
12.4.2 按一定比例扩大绘图边界 204
12.5 树数据相关工具 206
12.5.1 筛选树数据 206
12.5.2 展开嵌套的树数据 207
12.6 树相关工具 208
12.6.1 提取叶节点顺序 208
12.6.2 在分类单元标签前添加填充字符 210
12.7 交互式 ggtree 注释 211
12.8 本章练习题 211

第13章 可重复示例图库 213
13.1 绘制系统发育树与核苷酸序列之间的距离 213
13.2 以不同的符号点呈现自举值 217
13.3 突出显示不同分组 219
13.4 含有基因组位点结构信息的系统发育树 222

参考文献 223

附录 A 常见问题 224
A.1 安装相关问题 224
A.2 R 语言相关问题 225

- A.3 美学映射相关问题 225
  - A.3.1 美学映射的继承 225
  - A.3.2 切忌在美学映射中使用 "$" 226
- A.4 文本和标签相关问题 226
  - A.4.1 叶节点标签被截断 226
  - A.4.2 修改叶节点标签 227
  - A.4.3 修改叶节点标签格式 229
  - A.4.4 避免文本标签重叠 230
  - A.4.5 Newick 格式中的自举值 231
- A.5 分支设置 232
  - A.5.1 绘制与 plot.phylo() 函数效果相同的树 232
  - A.5.2 指定叶节点的顺序 233
  - A.5.3 缩短外群长分支 233
  - A.5.4 为树添加新的叶节点 234
  - A.5.5 更改任意分支的颜色或线条类型 236
  - A.5.6 在分支的任意位置添加符号点 236
- A.6 为不同的分面面板设置不同的 x 轴标签 237
- A.7 在树的底部图层绘制图形 239
- A.8 扩大环形布局或扇形布局树的内部空间 239
- A.9 使用离根最远的叶节点作为时间尺度树的原点 240
- A.10 删除环形布局树的空白边距 241
- A.11 编辑树图的细节 242
- 参考文献 242

附录 B 相关工具 243

- B.1 MircrobiotaProcess 包：将物种分类表转换为 treedata 对象 243
- B.2 rtol 包：Open Tree API 的 R 接口 244
- B.3 将 ggtree 对象转换为 plotly 对象 245
- B.4 绘制漫画风格的系统发育树（类似 xkcd） 246
- B.5 绘制 ASCII Art 形式的有根树 247
- B.6 放大树的选定部分 249

B.7 在 ggtree 包中使用 ggimage 包的提示 250
    B.7.1 示例 1：移除图像背景 250
    B.7.2 示例 2：在背景图像上绘制树 251
B.8 在 Jupyter Notebook 中运行 ggtree 包 251
参考文献 252

附录 C　练习题答案 253

第1篇
树数据的输入/输出及操作

第 1 章　导入带有数据的树文件

1.1　系统发育树构建概述

系统发育树（Phylogenetic Tree，简称"进化树"）是基于生物的遗传序列构建的，常用来描述生物群体之间的谱系关系。我们常用有根树来表示进化历史模型。树由树节点之间的"祖先—后代"关系，以及不同亲缘水平的"姐妹"或"表亲"生物的聚类所描绘，如图 1.1 所示。在传染病研究中，进化树通常由病原体的基因序列或基因组序列构建。我们可以通过找出哪个病原体样本在遗传上更接近另一个样本，从而更加深入地了解平时难以观察到的流行病学联系和流行病爆发的潜在源头。

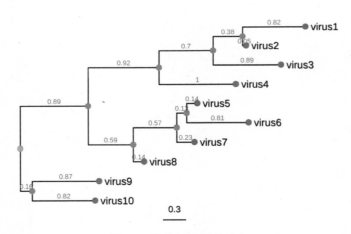

图 1.1　进化树的组成部分

外部节点（绿色圆点）又被称为"叶节点"，表示采样及测序的实际生物体（例如，传染病研究中的病毒），在进化生物学术语中又被称为"分类单元"。内部节点（蓝色圆点）表示外部节点的假设祖先。根（红色圆点）是进化树中所有物种的共同祖先。水平线条表示树的分支，又表示生物所发生的以时间或遗传分歧衡量的演变（灰色数字）。底部的线条表示这些分支长度的标尺。

基于基因序列，进化树可以通过两种方法来构建，一种是基于距离的方法；另一种是基于字符的方法。基于距离的方法是指基于所计算出的序列间遗传距离矩阵来构建进化树，其中包括非加权分组平均法（UPGMA）和邻接法（NJ）。基于字符的方法是指基于描述遗传字符演化的数学模型构建进化树，并根据其最优选择标准来找出最佳进化树，其中包括最大简约法（MP）[1]、最大似然法（ML）[2]和贝叶斯马尔科夫链蒙特卡洛法（BMCMC）[3]。

最大简约法（Maximum Parsimony，MP）的核心思想是物种在进化过程中会尽可能少地发生变化，并且以此为依据最小化序列字符改变的数量（例如，DNA 碱基替换的数量）。这种思想有点类似于奥卡姆剃刀原则，即可以解释数据的最简单假设就是最佳假设。未加权简约法认为不同序列字符（核苷酸或氨基酸）的突变具有同等的可能性，而加权简约法认为不同序列字符的突变的可能性不相等［例如，第三个密码子位置比其他密码子位置更多变；并且转换（transition）的频率高于颠换（transversion）的频率］。MP 方法直观而又简单，这使得它受到众多生物学家青睐。因为相较于相关分析的计算细节，学者们更专注于对问题的研究。但是，MP 方法也存在一些缺点，特别是对于进化树的推断来说，其结果可能因为存在长枝吸引效应（Long-Branch Attraction，LBA）而使推断结果产生偏差，如将远亲谱系错误地推断为近亲[4]。这是因为 MP 方法没有考虑到许多序列的进化因素［如逆转（reversals）和收敛（convergence）］，而这些因素在现有遗传数据中是很难被观察到的。

最大似然法（Maximum Likelihood，ML）和贝叶斯马尔科夫链蒙特卡洛法（Bayesian Markov Chain Monte Carlo，BMCMC）是构建进化树最常用的两种方法，并且被广泛应用于科研中。ML 方法和 BMCMC 方法都需要先确定序列演化的替代模型。不同的序列数据具有不同的替代模型，用以构建 DNA、密码子或氨基酸的演化过程。例如，存在着 JC69、K2P、F81、HKY 和 GTR[5]等多种核苷酸替代模型。这些模型可以与位点间不同的演化速率(记为 $+\Gamma$)[6]及恒定位点的比例(记为 $+I$) [7]一起使用。已有研究[8]表明，错误地指定替代模型可能会使系统发育推断产生偏差。所以我们建议先进行相关的检验以选出最合适的替代模型。

通过 ML 方法构建树的最优准则是找到由序列数据构建的似然值最大的树。ML 方法的过程很简单，计算一棵树的似然性并优化其拓扑和分支，直至找到最优树。PhyML 和 RAxML 中所应用的启发式搜索算法（Heuristic Search）通常用于根据似然标准找到最优树。贝叶斯方法基于给定的替代模型通过 MCMC 方法对树进行抽样，从而找到后验概率最大化的树。BMCMC 方法的优点之一是在抽样过程中，可以方便而自然地获得参数方差和树拓扑不确定性，包括进化枝后验概率。此外，拓扑不确定性对其他参数估计的影响也会被自然地整合到 BMCMC 系统发育框架中。

在相对简单的系统发育树中，与树的分支或节点相关的数据可以是枝长（表示遗传或时间分歧）和谱系支持度，例如，通过自举程序估算的自举值或在 BMCMC 方法分析中由抽样得出的进化枝后验概率等。

需要注意的是，BMCMC 方法是在 MCMC 方法基本上进行的推断，MCMC 是一种抽样方法，用于简化贝叶斯推断的计算方法。

1.2 系统发育树文件格式

用来存储系统发育树及其节点和分支相关数据的文件格式有很多种，其中有 3 种较为常见的文件格式，即 Newick[①]、NEXUS[9] 和 Phylip[10]。有些格式（如 NHX 格式）是从 Newick 格式扩展而来的。在进化生物学中，大多数软件都支持 Newick 和 NEXUS 作为输入格式，而有些软件（如 BEAST 和 MrBayes）则通过引入的新规则或数据模块来存储进化推论，从而更新标准文件格式。也有一些软件（如 PAML 和 r8s）所输出的日志文件只能被软件自身识别。

1.2.1 Newick 树文件

Newick 格式是计算机可读形式的树文件标准格式。图 1.2 所示为使用 Newick 文本编码树结构，叶节点于右侧对齐，在每个分支的中间标记分支的长度。

图 1.2 中的有根树可以表示为以下字符序列组成的 Newick 树文本格式。

((t2:0.04,t1:0.34):0.89,(t5:0.37,(t4:0.03,t3:0.67):0.9):0.59);

树文本以分号结尾，其内部节点由一对匹配的括号表示，而括号里面是该节

---

① Newick 官方文档请参见"外链资源"文档中第 1 章第 1 条

点的后代节点。例如，(t2:0.04,t1:0.34) 表示为 t2 和 t1 的父节点，t2、t1 均为其直系后代。同级的内部节点用逗号分隔，同时叶节点则用它们的名称表示。枝长（从父节点到子节点）由子节点后面的实数表示，两者之间用冒号分隔。与内部节点或分支相关联的单一数据（如自举值）可以被编码为节点标签，并用冒号前面简单的文本或数字表示。

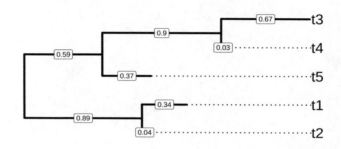

图 1.2　使用 Newick 文本编码树结构的示例

1984 年，Meacham 为 Phylip[11] 软件包开发了 Newick 格式。Newick 是当下使用最为广泛的进化树格式，被应用于许多程序中，如 Phylip、PAUP*[12]、TREE-PUZZLE[13]、MrBayes。Phylip 格式中包含了 Phylip 多序列比对结果和基于此比对结果所构建的 Newick 树文本。

1.2.2　NEXUS 格式

NEXUS 格式[9] 整合了 Newick 格式的树文本及相关信息，将它们分配到不同模块（blocks）中。NEXUS 模块具有以下结构。

```
#NEXUS
...
BEGIN characters;
...
END;
```

例如，可以将上面的示例树保存为以下 NEXUS 格式。

```
#NEXUS
[R-package APE, Fri Oct 29 10:37:40 2021]
```

```
BEGIN TAXA;
    DIMENSIONS NTAX = 5;
    TAXLABELS
        t2
        t1
        t5
        t4
        t3
    ;
END;
BEGIN TREES;
    TRANSLATE
        1   t2,
        2   t1,
        3   t5,
        4   t4,
        5   t3
    ;
    TREE * UNTITLED = [&R] ((1:0.04,2:0.34):0.89,(3:0.37,(4:0.03,5:0.67):0.9):0.59);
END;
```

树的注解信息（Comments）会被存储在方括号中。一些模块［如 TAXA（包含分类单元信息）、DATA（包含数据矩阵，如序列比对信息）和 TREES（包含进化树，即 Newick 树文本）］可以被大多数程序识别。由于模块非常多样化，因此有一些模块只能被特定程序识别。例如，由 PAUP* 输出的 Nexus 文件中，包含 PAUP* 命令的 paup 模块；以及 FigTree 输出的 Nexus 文件中，包含可视化设置的 figtree 模块等。NEXUS 格式用于将不同类型的数据分别存储到不同的模块中。支持读取 Nexus 文件的程序可以只解析它们能识别的模块，同时忽略它们不能识别的模块。这是一种很好的机制，因为它能使不同的程序解析相同的文件格式，不会由于文件中存在不受支持的数据类型而造成解析过程中断。但是，大多数程序仅支持解析 TAXA、DATA 和 TREES 模块。因此，开发一款能够读取 Nexus 文件中的各种数据模块的程序或平台对于系统发育数据的整合是非常有用的。

DATA 模块常用于存储序列比对结果。用户也可以利用 Phylip 格式来存储树和序列数据，其本质上就是 Phylip 多序列比对结果加上 Newick 树文本。

1.2.3　NHX 格式

Newick、NEXUS 和 Phylip 主要用于存储进化树和一些基本的内部节点或分支关联的单一数据。而基于 Newick（又被称为 New Hampshire）的 NHX 格式（New Hampshire eXtended），除了可以用于储存注释节点或分支中的单一数据，还可以通过引入 NHX 标签（tag）来将多个数据与树节点相关联（内/外部节点均可）。NHX 标签必须被放置于枝长之后的"[&&NHX"和"]"之间，这样就可以做到与 NEXUS 格式兼容，因为 NEXUS 会将"["和"]"之间的字符定义为注解信息。NHX 也是 PHYLDOG[14] 和 RevBayes[15] 的输出格式。ATV（A Tree Viewer）[16] 是一个 Java 工具，用于查看以 NHX 格式存储的注释数据，但此工具已经停止维护。

下面是 NHX 定义文档中的示例树。

```
(((ADH2:0.1[&&NHX:S=human], ADH1:0.11[&&NHX:S=human]):0.05
[&&NHX:S=primates:D=Y:B=100],ADHY:0.1[&&NHX:S=nematode],
ADHX:0.12[&&NHX:S=insect]):0.1[&&NHX:S=metazoa:D=N],(ADH4:0.09
[&&NHX:S=yeast],ADH3:0.13[&&NHX:S=yeast],ADH2:0.12[&&NHX:S=yeast],
ADH1:0.11[&&NHX:S=yeast]):0.1[&&NHX:S=Fungi])[&&NHX:D=N];
```

1.2.4　Jplace 格式

Matsen[17] 为存储映射至进化树的 NGS（二代测序）短读序列数据（用于宏基因组分类）设计了 Jplace 格式。Jplace 格式基于 JSON 格式构建，其中包含 5 个键（key）：树（tree）、域（fields）、放置信息（placements）、元数据（metadata）和版本（version）。其中，树（tree）包含从 Newick 树格式扩展的树文本，它将边（edge）标签放在枝长信息之后的方括号中（如果其包含分支标签信息），并将边编号（edge number）放在边标签之后的花括号中。域（fields）中包含所放数据的头信息（header information）。放置信息（placements）是一个由多个 pquery 组成的列表。每个 pquery 包含两个键，一个是 p，另一个是 n（或 nm）。其中，p 表示具体的放置信息，n 表示名称（或 nm 表示多个名称）。p 是由该 pquery 中的放置信息组成的列表。

下面是一个 Jplace 格式的示例。

```
{
    "tree": "((((((((A:4{1},B:4{2}):6{3},C:5{4}):8{5},D:6{6}):
    3{7},E:21{8}):10{9},((F:4{10},G:12{11}):14{12},H:8{13}):
    13{14}):13{15},((I:5{16},J:2{17}):30{18},(K:11{19},
    L:11{20}):2{21}):17{22}):4{23},M:56{24});",
    "placements": [
    {"p":[24, -61371.300778, 0.333344, 0.000003, 0.003887],
     "n":["AA"]
    },
    {"p":[[1, -61312.210786, 0.333335, 0.000001, 0.000003],
          [2, -61322.210823, 0.333322, 0.000003, 0.000003],
          [3, -61352.210823, 0.333322, 0.000961, 0.000003]],
     "n":["BB"]
    },
    {"p":[[8, -61312.229128, 0.200011, 0.000001, 0.000003],
          [9, -61322.229179, 0.200000, 0.000003, 0.000003],
          [10, -61342.229223, 0.199992, 0.000003, 0.000003]],
    "n":["CC"]
    }
    ],
    "metadata": {"info": "a jplace sample file"},
    "version" : 2,
    "fields": ["edge_num", "likelihood", "like_weight_ratio",
    "distal_length", "pendant_length"]
    ]
}
```

PPLACER[18] 和 EPA（Envolutionary Placement Algorithom）[19] 都是以 Jplace 格式输出结果的。但是这两个程序并不包含对进化放置信息进行可视化的工具。PPLACER 中的 placeviz 用于将 Jplace 格式的文件转换为 phyloXML 或 Newick 格式的文件，并使用 Archaeopteryx 进行可视化。

1.2.5 利用软件输出文件

RAxML[20] 通过存储自举值（bootstrap）为内部节点标签，便可将其输出为 Newick 格式的文件。RAxML 的另一种输出格式，是将自举值放在枝长后的方括号内。大多数支持解析 Newick 格式的软件都无法支持解析此种以枝长后的方括号存储自举值的格式，它们会忽略方括号中所包含的内容。

BEAST[21] 用于输出 NEXUS 格式的文件，它也同样在树模块中引入方括号来存储经 BEAST 推断所得的进化证据。如果方括号内特征值的长度大于 1（例如，最高概率密度或替换速率的范围），则还会在其方括号内添加花括号来存储这些信息。这些括号被放置在节点和枝长之间（在标签与冒号之间）。由于 Newick 格式中并没有定义方括号，且方括号在 NEXUS 格式中也是作为注解信息的保留字符存在的，因此标准的 NEXUS 解析器会将这些信息忽略。

下面是 BEAST 输出的 TREE 模块的示例。

```
TREE * TREE1 = [&R] (((11[&length=9.47]:9.39,14[&length=6.47]:6.39)
[&length=25.72]:25.44,4[&length=9.14]:8.82)[&length=3.01]:3.1,
(12[&length=0.62]:0.57,(10[&length=1.6]:1.56,(7[&length=5.21]:5.19,
((((2[&length=3.3]:3.26,(1[&length=1.34]:1.32,(6[&length=0.85]:0.83,
13[&length=0.85]:0.83)[&length=2.5]:2.49)[&length=0.97]:0.94)
[&length=0.5]:0.5,9[&length=1.76]:1.76)[&length=2.41]:2.36,
8[&length=2.19]:2.11)[&length=0.27]:0.24,(3[&length=3.33]:3.31,
(15[&length=5.29]:5.27,5[&length=3.29]:3.27)[&length=1.04]:1.04)
[&length=1.98]:2.04)[&length=2.83]:2.84)[&length=5.39]:5.37)
[&length=2.02]:2)[&length=4.35]:4.36)[&length=0];
```

BEAST 会根据在 BEAUti 中所使用的分析模型，输出包含相应的进化推论的文件。例如，分子钟分析中包含了 rate、length、height、posterior，以及对应的用于不确定性估计的 HPD 与极差。其中，rate 表示估计的分支的进化速率，length 表示以年为单位的枝长，height 表示从节点到根的时间，posterior 表示贝叶斯进化枝的可信度值。上面示例是通过分子钟分析输出的树文件，其中就应该包含以上这些推论，但为了节省空间，上面示例只显示了所估算的枝长 length。此外，MEGA（Molecular Evolutionary Genetics Analysis）[22] 也支持以 BEAST 兼容的 NEXUS 格式导出进化树。

MrBayes[23] 是一个使用 MCMC 方法进行后验概率分布抽样的软件。利用该软件输出文件可通过两组方括号来对节点和分支分别进行注释。下面示例分别注释了节点的进化枝后验概率与分支的枝长估计值。

```
tree con_all_compat = [&U] (8[&prob=1.0]:2.94e-1[&length_
mean=2.9e-1],
10[&prob=1.0]:2.25e-1[&length_mean=2.2e-1],((((1[&prob=1.0]:1.43e-1
[&length_mean=1.4e-1],2[&prob=1.0]:1.92e-1[&length_mean=1.9e-1])
[&prob=1.0]:
```

```
1.24e-1[&length_mean=1.2e-1],9[&prob=1.0]:2.27e-1[&length_
mean=2.2e-1])
[&prob=1.0]:1.72e-1[&length_mean=1.7e-1],12[&prob=1.0]:5.11e-1
[&length_mean=5.1e-1])[&prob=1.0]:1.76e-1[&length_mean=1.7e-1],
(((3[&prob=1.0]:5.46e-2[&length_mean=5.4e-2],(6[&prob=1.0]:1.03e-2
[&length_mean=1.0e-2],7[&prob=1.0]:7.13e-3[&length_mean=7.2e-3])
[&prob=1.0]:
6.93e-2[&length_mean=6.9e-2])[&prob=1.0]:6.03e-2[&length_
mean=6.0e-2],
(4[&prob=1.0]:6.27e-2[&length_mean=6.2e-2],5[&prob=1.0]:6.31e-2
[&length_mean=6.3e-2])[&prob=1.0]:6.07e-2[&length_mean=6.0e-2])
[&prob=1.0]:,
1.80e-1[&length_mean=1.8e-1]11[&prob=1.0]:2.37e-1[&length_
mean=2.3e-1])
[&prob=1.0]:4.05e-1[&length_mean=4.0e-1])[&prob=1.0]:1.16e+000
[&length_mean=1.162699558201079e+000])[&prob=1.0][&length_mean=0];
```

为了节省空间,此示例中的大部分推论信息都被删除,只保留了表示进化枝概率的 prob 和表示枝长平均值的 length_mean。此文件的完整版本还包含了表示概率推断的 prob_stddev、prob_range、prob(percent)、prob+-sd,以及每个分支的 length_median、length_95%_HPD。

大部分支持解析 NEXUS 格式的软件在解析 BEAST 和 MrBayes 的输出文件时,都会将其中的进化推论当作注解信息而忽略。虽然 FigTree 支持解析 BEAST 与 MrBayes 输出文件中的进化推论,并可以使用它们在可视化层面对进化树进行注释,但想要以此将这些数据提取出来进行进一步分析仍然非常困难。

HyPhy[24] 可用于系统发育分析,如祖先序列重建等。在进行祖先序列重建时,HyPhy 会将包含所构建的祖先序列及 Newick 树文本的 NEXUS 格式文件作为主要的输出对象。但输出的文件并没有完全遵循 NEXUS 格式的标准。其中 TAXA 模块只储存了祖先节点的标签,而不含有外部节点的标签。MATRIX 模块只储存了祖先节点的序列比对结果。由于其中不含有节点标签,因此无法被重新指定回存储于 TREES 模块中的树。以下为 HyPhy 的示例输出(为了节省空间,只显示了序列比对的前 72bp)。

```
#NEXUS

[
Generated by HYPHY 2.0020110620beta(MP) for MacOS(Universal Binary)
    on Tue Dec 23 13:52:34 2014
```

]

BEGIN TAXA;
    DIMENSIONS NTAX = 13;
    TAXLABELS
        'Node1' 'Node2' 'Node3' 'Node4' 'Node5' 'Node12' 'Node13' 'Node15'
            'Node18' 'Node20' 'Node22' 'Node24' 'Node26' ;
END;

BEGIN CHARACTERS;
    DIMENSIONS NCHAR = 2148;
    FORMAT
        DATATYPE = DNA

        GAP=-
        MISSING=?
        NOLABELS
    ;

MATRIX

ATGGAAGACTTTGTGCGACAATGCTTCAATCCAATGATCGTCGAGCTTGCGGAAAAGGCAATGAAAGAATAT
ATGGAAGACTTTGTGCGACAATGCTTCAATCCAATGATCGTCGAGCTTGCGGAAAAGGCAATGAAAGAATAT
ATGGAAGACTTTGTGCGACAATGCTTCAATCCAATGATCGTCGAGCTTGCGGAAAAGGCAATGAAAGAATAT
ATGGAAGACTTTGTGCGACAATGCTTCAATCCAATGATCGTCGAGCTTGCGGAAAAGGCAATGAAAGAATAT
ATGGAAGACTTTGTGCGACAATGCTTCAATCCAATGATTGTCGAGCTTGCGGAAAAGGCAATGAAAGAATAT
ATGGAAGACTTTGTGCGACAATGCTTCAATCCAATGATCGTCGAGCTTGCGGAAAAGGCAATGAAAGAATAT
ATGGAAGACTTTGTGCGACAATGCTTCAATCCAATGATCGTCGAGCTTGCGGAAAAGGCAATGAAAGAATAT
ATGGAAGACTTTGTGCGACAATGCTTCAATCCAATGATCGTCGAGCTTGCGGAAAAGGCAATGAAAGAATAT
ATGGAAGACTTTGTGCGACAATGCTTCAATCCAATGATCGTCGAGCTTGCGGAAAAGGCAATGAAAGAATAT
ATGGAAGACTTTGTGCGACAATGCTTCAATCCAATGATCGTCGAGCTTGCGGAAAAGGCAATGAAAGAATAT
ATGGAAGACTTTGTGCGACAATGCTTCAATCCAATGATCGTCGAGCTTGCGGAAAAGGCAATGAAAGAATAT
ATGGAAGACTTTGTGCGACAATGCTTCAATCCAATGATCGTCGAGCTTGCGGAAAAGGCAATGAAAGAATAT
ATGGAAGACTTTGTGCGACAGTGCTTCAATCCAATGATCGTCGAGCTTGCGGAAAAGGCAATGAAAGAATAT
END;

BEGIN TREES;
    TREE tree = (K,N,(D,(L,(J,(G,((C,(E,O)),(H,(I,(B,(A,(F,M)))))))))));
END;

其他应用程序也可以输出有关联数据的进化树并含有丰富信息的文本。例如，通过 r8s[25] 会在日志文件中输出 3 棵树，即分别以时间、替换率和绝对替换率作为枝长标尺的 TREE、RATE 和 PHYLO。

PAML（Phylogenetic Analysis by Maximum Likelihood）[26] 是一个用于 DNA 或蛋白质序列系统发育分析的程序。BASEML 和 CODEML 作为两个主要的程序，实现了多种模型在系统发育分析中的应用。BASEML 可以通过许多可用的核苷酸替换模型（包括 JC69、K80、F81、F84、HKY85、T92、TN93 和 GTR）来估计树拓扑结构、枝长和替换参数。CODEML 可以通过密码子替换模型[27] 估计同义和非同义替换发生的速率，以及进行正向选择的似然比检验。

BASEML 用于输出 mlb 格式的文件，其中包含输入序列（分类单元）的比对结果、枝长信息的进化树，以及替代模型和其他所估算出的参数。同时，BASEML 还用于输出一个 rst 补充结果文件，其中包含序列比对结果（如果进行了祖先序列重建，则会包含祖先序列）和序列比对中每个位点进化的朴素经验贝叶斯（Naive Empirical Bayes，NBE）概率。CODEML 用于输出 mlc 格式的文件，其中包含树结构，以及所估算出的同义替代速率和非同义替代速率。与 BASEML 类似，CODEML 也用于输出一个 rst 补充结果文件，不同之处在于其中位点被定义为密码子而不是核苷酸。解析这些文件的过程十分乏味，而且经常需要对数据进行许多后期处理，也需要用户拥有编程方面的专业知识（如 Python① 或 Perl②）。

有很多进化树的文件都通过引入方括号来存储额外信息。例如，RAxML 格式用方括号存储自举值、NHX 格式用方括号存储注释、Jplace 格式用方括号存储边标签、BEAST 格式用方括号存储进化估计等。但是不同软件中方括号的位置并不一致，而且不同软件会使用不同的规则来将相关数据存储到方括号中，这就使得解析相关数据变得非常困难。大多数软件只支持解析特定兼容文件中树的拓扑结构，而忽略了方括号内存储的信息。同时，对于一些包含无效字符的文件格式（如 Jplace 格式的 tree 域中的花括号），标准的解析器无法将树的拓扑结构解析出来。

由于不同的进化推理软件会产生不同的输出文件格式，所以我们很难提取出

---

① 在 Python 中解析 PAML 输出的详细内容请参见"外链资源"文档中第 1 章第 2 条
② 在 Perl 中解析 PAML 输出的详细内容请参见"外链资源"文档中第 1 章第 3 条

所需的系统发育信息或分类单元相关信息，并将其显示到同一棵进化树上进行进一步分析。虽然 FigTree 支持输出 BEAST 格式，但不支持大多数其他软件所输出的含有进化推理或关联数据的文件格式。而对于那些含有丰富输出的文本文件（如 r8s、PAML 等），任何用以查看进化树的软件都无法解析其中的树结构。用户需要具有很强的专业知识才能从输出文件中手动提取出进化树及其他所需的结果数据。同时，这种手动操作不仅十分缓慢，而且非常容易出错。

我们很难从这些在系统发育分析领域常用软件的不同输出格式中获取含有进化数据的系统发育树。有些输出格式（如 PAML 和 Jplace）缺少支持解析这些文件的软件或编程库，而其他输出格式（如 BEAST 和 MrBayes）只能解析出树结构而不能解析出其中的进化推论，因为大多数软件会将这些存储于方括号内的信息当作注解信息而省略。虽然 FigTree 能对 BEAST 和 MrBayes 所推断的进化统计数据结果进行可视化，但是不支持将这些数据提取出来以进行进一步的分析。不同的软件包实现不同的算法用于不同的分析 [例如，用于 dN/dS 选择压力分析的 PAML、用于祖先序列分析的 HyPhy 和用于天际线分析（skyline analysis）的 BEAST]。因此，在进行基因组序列数据分析时，我们需要一个能高效灵活地整合不同的分析推断结果的工具，以便于更加全面地理解、比较数据，并进行进一步的分析。这也是我们想要开发一个编程函数库以解析来自各种来源的进化树及数据的动力所在。

1.3 使用 treeio 导入树及相关数据

我们经常使用进化树来展现物种间的进化关系，而与树中的物种、菌株等分类单元关联的信息则可以在进化树提供的进化历史背景下进行进一步的分析。例如，我们可以通过研究树中不同流感病毒株的宿主信息来探究某一病毒谱系的宿主范围。除此之外，诸如分离株宿主、感染时间、感染地点等与病毒株直接相关的元数据也常被应用于系统发育的进化分析及比较分析的建模中，以推断它们在病毒进化或传播过程中的动力学关系。我们通常将这类的元数据、表型数据或实验数据存储为与节点或分支关联的注释数据，使用不同的软件来存储此类树+注释的文件格式往往是不一致的。

目前来说，将进化树导入 R 还存在着许多限制。Newick 和 NEXUS 格式尚可以被 ape、phylobase、NeXML 等多种包导入，NeXML 格式可以使用 RNeXML 包来解析。系统发育分析领域中其他的一些常用软件的分析结果却没有得到很好的支持。例如，PHYLOCH 包虽然支持导入 BEAST 及 MrBayes 的输出结果，但是只能解析内部节点属性，而忽略叶节点属性[①]。还有很多软件所输出的进化树及关联数据仍需要熟悉编程的人员才能将其导入 R。除此之外，将一些外部数据（如实验数据及临床数据等）与进化树关联起来也是一件难事。

为了填补 R[28] 社区在这方面的空白，我们基于 R 语言开发了一款名为 treeio[29] 的 R 包，而它能解析大多数常用进化分析软件输出的不同格式的进化树文件。treeio 能将进化树的拓扑结构及其关联的多种数据和进化推论一并解析，如 NHX 注释、BEAST 或 r8s[25] 的分子钟速率推断、CODEML 推断的同义替换与非同义替换、BASEML、HyPhy 及 CODEML 重构的祖先序列等。我们在 treeio[29] 中开发了多种解析函数（见表 1.1）来支持读取这些文件格式，目前已经支持读取的文件格式包括 Newick、NEXUS、NHX、Jplace 及 Phylip，也支持读取 ASTRAL、BEAST、EPA、HyPhy、MEGA、MrBayes、PAML、PHYLDOG、PPLACER、r8s、RAxML、RevBayes 等多种软件的输出结果。

表 1.1　解析函数名称及其说明

解析函数名称	说　　明
read.astral()	解析 ASTRAL 的输出文件
read.beast()	解析 BEAST 的输出文件
read.codeml()	解析 CODEML 的输出文件（rst 文件及 mlc 文件）
read.codeml_mlc()	解析 mlc 文件
read.fasta()	解析 FASTA 格式的序列文件
read.hyphy()	解析 HYPHY 的输出文件
read.hyphy.seq()	解析 HYPHY 输出文件中的祖先序列
read.iqtree()	解析 IQ-Tree Newick 字符串，也支持解析 Iqtree 的 SH-aLRT 及 UFBoot 支持度
read.jplace()	解析由 EPA 及 pplacer 输出的 Jplace 文件
read.jtree()	解析 jtree 文件格式
read.mega()	解析 MEGA 输出的 Nexus 文件

① 相关讨论请参见"外链资源"文档中第 1 章第 4 条

续表

解析函数名称	说明
read.mega_tabular()	解析 MEGA 输出的 Tabular 文件
read.mrbayes()	解析 MrBayes 的输出文件
read.newick()	解析 Newick 字符串,且支持将节点标签解析为支持度
read.nexus()	解析标准的 Nexus 文件(该函数是从 ape 包重新导出的)
read.nhx()	解析 PHYLDOG 及 RevBayes 所输出的 NHX 文件
read.paml_rst()	解析 BaseML 或 CODEML 输出的 rst 文件
read.phylip()	解析 phylip 文件(其中包含 phylip 序列比对结果与 Newick 字符串)
read.phylip.seq()	从 phylip 文件中解析多序列比对结果
read.phylip.tree()	从 phylip 文件中解析 Newick 字符串
read.phyloxml()	解析 phyloXML 文件
read.r8s()	解析 r8s 的输出文件
read.raxml()	解析 RaxML 的输出文件
read.tree()	解析 Newick 字符串(该函数是从 ape 包重新导出的)

treeio 包为 R 中进化树的输入与输出定义了多种基本的类与函数。我们可以利用这些类与函数,将各类不同软件所推断的进化证据解析并导入 R 中。例如,CODEML[26] 推断出的祖先序列或计算出的 dn/ds 值、BEAST[21] 推断出的进化枝支持度(后验概率)、EPA[19] 和 PPLACER[18] 推断出的短读序列放置信息等,这些进化证据都可以通过 treeio 导入 R 以供后续的各类分析并通过 ggtree 包[30] 实现对进化树可视化上的注释。随着各类分析工具及模型的发展,将多种对相同进化树在不同软件分析中的结果整合起来共同分析也变得越来越困难。为了解决这个问题,treeio 包[29] 设计了 merge_tree() 函数来整合不同来源的树数据。除此之外,treeio 包还支持将外部数据与进化树关联到一起。

在解析完树后,treeio 包使用在 tidytree 包中定义的 S4 类——treedata 用来存储进化树及其相关数据。这些相关数据在 treedata 中会被映射到分支或节点上,以便后续能更有效地使用 ggtree[30] 和 ggtreeExtra[31] 来对树及其注释信息进行可视化。像这样一个专门为解析、整合及注释系统发育学数据而设计的可编程平台可以让我们更加容易地发现进化动力学关系及相关性模式。

图 1.3 所示为 treeio 包[29] 概述及其与 tidytree 和 ggtree 的关系。treeio 包用于

解析多种文件格式、软件输出文件中的进化树及关联数据。treeio 包使用 treedata 对象来存储系统发育树及其分支、节点关联的数据，也提供了多种函数来操作带有数据的树文件。用户可以将 treedata 对象转换为整洁数据框（tidy data frame，其中每一行表示树中的一个节点，每一列表示一个变量），从而使用 tidytree 中的整洁接口（tidy interface）来对树进行处理；同时，可以选择将树的拓扑结构从 treedata 中提取出来，存储到 Newick 文件或 Nexus 文件中，或者将树及关联数据一起导入同一个文件中（BEAST、Nexus 或 jtree），这些树关联的数据可以在可视化中使用 ggtree 对树进行注释。除了常见进化树的可视化，ggtree 也支持其他树形对象的可视化，包括处理宏基因组学数据的 phyloseq 对象及处理病毒爆发数据的 obkData 对象。ggtree 还支持对 phylo、multiPhylo（ape 包）、phylo4、phylo4d（phylobase 包）、phylog（ade4 包）、phyloseq（phyloseq 包）obkData（OutbreakTools 包）等多种 R 社区中所定义的用以存储树与特定领域数据的树对象的可视化，以及层次聚类结果（如 hclust 与 dendrogram 对象）的可视化。

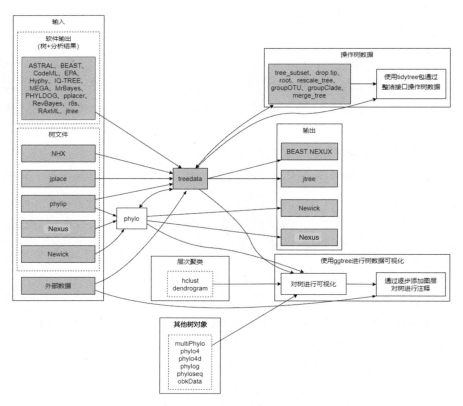

图 1.3　treeio 包概述及其与 tidytree 和 ggtree 的关系

1.3.1 treeio 简介

treeio 包[29]针对进化树的输入和输出进行设计，定义了 S4 类对象 treedata 来存储进化树及各种软件包输出在内的不同种类的关联数据或协变量，同时设计了相应的解析器来将对应的进化树及其注释数据转换为 R 的对象，以便于后续的数据处理及分析。除此之外，treeio 还设计了多种存取器（accessor），便于提取树文件中的注释信息。例如，get.fields() 函数用于识别并提取树中所能获取到的注释信息；get.placements() 函数用于提取 PPLACER、EPA 软件所得出的进化放置信息（phylogenetic placement）；get.subs() 函数用于提取从父节点到子节点的遗传替换信息；get.tips() 函数用于提取叶节点的序列信息。

目前，R 社区内大部分进化树相关的 R 包中使用的仍是 ape 包[32]中定义的 S3 类对象 phylo。为了利用 treeio 导入树后，能继续使用这些包进行分析，treeio 中的 as.phylo() 函数并将其生成的 S4 类对象 treedata 转换为 S3 类对象 phylo。但需要注意的是，转换后的 phylo 对象只含有进化树的拓扑结构信息，而不包含任何注释信息。treeio 中的 as.treedata() 函数用于将含有进化分析结果（如利用 ape 计算出的自举值或利用 phangorn[33]推断的祖先状态等）的 phylo 对象转换为 S4 类对象 treedata。转换之后，用户可以更加便利地把注释信息映射到树结构上，以便于后续使用 ggtree[30]实现对树的可视化。

为了便于将多种不同种类的数据整合到进化树，在 treeio[29]中定义了 merge_tree() 函数。而对于存储在用户自定义文件类型中的其他信息（如采样地点、分类学信息、实验结果、演化性状等进化证据），先通过 treeio 及 tidytree 中定义的 full_join() 函数对这些数据进行整合，再将这些文件通过 R 中的 I/O 函数读取后，我们便可以通过 full_join() 函数（或者使用 ggtree 中的 %<+% 操作符）将这些信息与树关联起来，并将它们变为分支或节点的属性，以便于后续与其他数据进行比对，或者在可视化时将其展示于树上[34]。

同时，为了更好地存储树及其关联的复杂数据，treeio 还提供了 write.beast() 函数与 write.jtree() 函数，用于将 treedata 对象输出至单一文件中。

1.3.2 treeio 解析函数演示

1.3.2.1 解析 BEAST 输出文件

```
file <- system.file("extdata/BEAST", "beast_mcc.tree",
package="treeio")
beast <- read.beast(file)
beast
```

```
'treedata' S4 object that stored information
## of
##   '/home/ygc/R/library/treeio/extdata/BEAST/beast_mcc.tree'.
##
## ...@ phylo:
##
## Phylogenetic tree with 15 tips and 14 internal nodes.
##
## Tip labels:
##   A_1995, B_1996, C_1995, D_1987, E_1996, F_1997, ...
##
## Rooted; includes branch lengths.
##
## with the following features available:
##   'height', 'height_0.95_HPD', 'height_median',
## 'height_range', 'length', 'length_0.95_HPD',
## 'length_median', 'length_range', 'posterior', 'rate',
## 'rate_0.95_HPD', 'rate_median', 'rate_range'.
```

由于 % 在 names 中并不是一个合法的字符，treeio 会自动将特征名中的百分数（x%）转换为小数形式（0.x），如将 length_95%_HPD 转换为 length_0.95_HPD。

同时，除了树结构本身，treeio 还能将所有 BEAST 软件的分析结果一起存储到 S4 对象中，以供在后续流程中提供对树的注释。

1.3.2.2 解析 MEGA 输出文件

MEGA（Molecular Evolutionary Genetics Analysis）[22] 是一款可以进行序列比对及进化树构建的软件。在构建进化树后，MEGA 支持将进化树输出为

Newick、Tabular 和 Nexus 这 3 种格式的文件。其中，treeio 中的 read.tree() 函数或 read.newick() 函数用于解析 Newick 文件；MEGA 与 BEAST 输出的都是非标准的 Nexus 文件。treeio[29] 中的 read.mega() 函数可用于解析 MEGA 输出的 Nexus 文件。

```
file <- system.file("extdata/MEGA7", "mtCDNA_timetree.nex",
                    package = "treeio")
read.mega(file)
```

```
## 'treedata' S4 object that stored information
## of
##   '/home/ygc/R/library/treeio/extdata/MEGA7/mtCDNA_timetree.nex'.
##
## ...@ phylo:
##
## Phylogenetic tree with 7 tips and 6 internal nodes.
##
## Tip labels:
##   homo_sapiens, chimpanzee, bonobo, gorilla,
## orangutan, sumatran, ...
##
## Rooted; includes branch lengths.
##
## with the following features available:
##   'branch_length', 'data_coverage', 'rate',
## 'reltime', 'reltime_0.95_CI', 'reltime_stderr'.
```

我们利用 MEGA 的 Tabular 输出文件将树及其关联信息（本示例中为分歧时间）存储在一个表格化的平面文本文件中。treeio 中的 read.mega_tubular() 函数用于解析 MEGA 输出的 Tabular 文件。

```
file <- system.file("extdata/MEGA7", "mtCDNA_timetree_tabular.txt",
                    package = "treeio")
read.mega_tabular(file)
```

```
## 'treedata' S4 object that stored information
## of
##   '/home/ygc/R/library/treeio/extdata/MEGA7/mtCDNA_timetree_tabular.txt'.
##
## ...@ phylo:
```

```
## 
## Phylogenetic tree with 7 tips and 6 internal nodes.
## 
## Tip labels:
##   chimpanzee, bonobo, homo sapiens, gorilla,
## orangutan, sumatran, ...
## Node labels:
##   , , demoLabel2, , ,
## 
## Rooted; no branch lengths.
## 
## with the following features available:
##   'RelTime', 'CI_Lower', 'CI_Upper', 'Rate', 'Data
## Coverage'
```

1.3.2.3 解析 MrBayes 输出文件

MrBayes 软件输出的同样是非标准 Nexus 文件，与 BEAST 的输出文件稍有不同但大致相似。因此，在使用 treeio 中的 read.mrbayes() 函数解析 MrBayes 输出文件时，会在内部调用 read.beast() 函数。

```
file <- system.file("extdata/MrBayes", "Gq_nxs.tre", package="treeio")
read.mrbayes(file)
```

```
## 'treedata' S4 object that stored information
## of
##   '/home/ygc/R/library/treeio/extdata/MrBayes/Gq_nxs.tre'.
## 
## ...@ phylo:
## 
## Phylogenetic tree with 12 tips and 10 internal nodes.
## 
## Tip labels:
##   B_h, B_s, G_d, G_k, G_q, G_s, ...
## 
## Unrooted; includes branch lengths.
## 
## with the following features available:
##   'length_0.95HPD', 'length_mean', 'length_median',
## 'prob', 'prob_range', 'prob_stddev', 'prob_percent',
## 'prob+-sd'.
```

1.3.2.4 解析 PAML 输出文件

PAML（Phylogenetic Analysis by Maximum Likelihood）是一个使用最大似然法来对 DNA 和蛋白质序列进行系统发育分析的软件包，在其子程序 BASEML 与 CODEML 中实现了多种树搜索算法。treeio 中的 read.paml_rst() 函数用于解析 BASEML 与 CODEML 输出的 rst 文件。两者生成的 rst 文件的主要区别在于序列间的空格位置不同。在 BASEML 生成的 rst 文件中，每 10 个碱基间会由一个空格隔开，而在 CODEML 生成的 rst 文件中则是每 3 个碱基（一个三联体）间会由一个空格隔开。

```
brstfile <- system.file("extdata/PAML_Baseml", "rst", package="treeio")
brst <- read.paml_rst(brstfile)
brst
```

```
## 'treedata' S4 object that stored information
## of
##   '/home/ygc/R/library/treeio/extdata/PAML_Baseml/rst'.
##
## ...@ phylo:
##
## Phylogenetic tree with 15 tips and 13 internal nodes.
##
## Tip labels:
##    A, B, C, D, E, F, ...
## Node labels:
##    16, 17, 18, 19, 20, 21, ...
##
## Unrooted; includes branch lengths.
##
## with the following features available:
##    'subs', 'AA_subs'.
```

也可以通过 read.paml_rst() 函数来解析 CODEML 中的 rst 文件。

```
crstfile <- system.file("extdata/PAML_Codeml", "rst", package="treeio")
## 此处 type 参数可以为 "Marginal" 或 "Joint"
crst <- read.paml_rst(crstfile, type = "Joint")
crst
```

```
## 'treedata' S4 object that stored information
```

```
## of
##      '/home/ygc/R/library/treeio/extdata/PAML_Codeml/rst'.
##
## ...@ phylo:
##
## Phylogenetic tree with 15 tips and 13 internal nodes.
##
## Tip labels:
##    A, B, C, D, E, F, ...
## Node labels:
##    16, 17, 18, 19, 20, 21, ...
##
## Unrooted; includes branch lengths.
##
## with the following features available:
##    'subs', 'AA_subs'.
```

BASEML 与 CODEML 通过边缘最大似然法或联合最大似然法重建的祖先序列会被存储在 S4 类对象 treedata 中，并映射到对应的节点上。treeio[29] 会自动确定每个分支两端节点上的序列间的替换信息。我们通过将核苷酸序列翻译为氨基酸序列，也可以确定氨基酸序列间的替换信息。这些经由计算的替换信息会被存储在 treedata 对象中，以便于后续对进化树进行注释与可视化。

CODEML 可以推断选择压力，并估算 dN/dS、dN 及 dS 的值。这些信息会被存储于 CODEML 输出的 mlc 文件中，并通过 read.codeml_mlc() 函数来解析。

```
mlcfile <- system.file("extdata/PAML_Codeml", "mlc", package="treeio")
mlc <- read.codeml_mlc(mlcfile)
mlc
```

```
## 'treedata' S4 object that stored information
## of
##      '/home/ygc/R/library/treeio/extdata/PAML_Codeml/mlc'.
##
## ...@ phylo:
##
## Phylogenetic tree with 15 tips and 13 internal nodes.
##
## Tip labels:
##    A, B, C, D, E, F, ...
```

```
## Node labels:
##  16, 17, 18, 19, 20, 21, ...
##
## Unrooted; includes branch lengths.
##
## with the following features available:
##  't', 'N', 'S', 'dN_vs_dS', 'dN', 'dS', 'N_x_dN',
##  'S_x_dS'.
```

treeio 不仅能单独解析 rst 文件与 mlc 文件，而且能使用 read.codeml() 函数同时读取 rst 文件和 mlc 文件。

```
## 我们可以通过 tree = "rst" 或 tree = "mlc"
## 来指定要使用哪个文件中的树作为本项目中的基础树

ml <- read.codeml(crstfile, mlcfile, tree = "mlc")
ml
```

```
## 'treedata' S4 object that stored information
## of
##  '/home/ygc/R/library/treeio/extdata/PAML_Codeml/rst',
##  '/home/ygc/R/library/treeio/extdata/PAML_Codeml/mlc'.
##
## ...@ phylo:
##
## Phylogenetic tree with 15 tips and 13 internal nodes.
##
## Tip labels:
##  A, B, C, D, E, F, ...
## Node labels:
##  16, 17, 18, 19, 20, 21, ...
##
## Unrooted; includes branch lengths.
##
## with the following features available:
##  'subs', 'AA_subs', 't', 'N', 'S', 'dN_vs_dS', 'dN',
##  'dS', 'N_x_dN', 'S_x_dS'.
```

所有 rst 文件与 mlc 文件中的特征数据都会被导入单一 S4 对象中，以便后续对进化树进行注释与可视化。例如，我们可以在同一棵进化树上[30]同时展示 mlc 文件中获取的 dN/dS 值及 rst 文件中获取的氨基酸替换信息。

1.3.2.5 解析 HyPhy 输出文件

HyPhy（Hypothesis testing using Phylogenies）是一个用于进行遗传序列分析的软件包。HyPhy 会将推断的祖先序列与树的拓扑结构一起输出到 Nexus 文件中，并使用 treeio 中的 read.hyphy.seq() 函数解析此文件。

```
ancseq <- system.file("extdata/HYPHY", "ancseq.nex", package="treeio")
read.hyphy.seq(ancseq)
```

```
## 13 DNA sequences in binary format stored in a list.
##
## All sequences of same length: 2148
##
## Labels:
## Node1
## Node2
## Node3
## Node4
## Node5
## Node12
## ...
##
## Base composition:
##     a     c     g     t
## 0.335 0.208 0.237 0.220
## (Total: 27.92 kb)
```

如果想要将序列映射至树上，则需要提供树内部节点的标签；如果想要确定替换信息，则需要提供叶节点的序列信息。在本示例中，与解析 CODEML 输出文件时类似，替换信息会被 treeio 自动确定。

```
nwk <- system.file("extdata/HYPHY", "labelledtree.tree",
package="treeio")
tipfas <- system.file("extdata", "pa.fas", package="treeio")
hy <- read.hyphy(nwk, ancseq, tipfas)
hy
```

```
## 'treedata' S4 object that stored information
## of
##  '/home/ygc/R/library/treeio/extdata/HYPHY/labelledtree.tree'.
##
```

```
## ...@ phylo:
##
## Phylogenetic tree with 15 tips and 13 internal nodes.
##
## Tip labels:
##   K, N, D, L, J, G, ...
## Node labels:
##   Node1, Node2, Node3, Node4, Node5, Node12, ...
##
## Unrooted; includes branch lengths.
##
## with the following features available:
##   'subs', 'AA_subs'.
```

1.3.2.6　解析 r8s 输出文件

r8s 包可以通过参数、半参数及非参数的方法来设定宽松分子钟，以便更加精准地估算分歧时间及进化速率[25]。在输出时，r8s 会将 3 棵进化树存储到一个日志文件中。这 3 棵进化树分别为时间树（TREE）、速率树（RATO）与绝对替换树（PHYLO）。

时间树使用分歧时间作为树枝长的标尺，速率树使用替换速率作为树枝长的标尺，绝对替换树以替换的绝对值作为树枝长的标尺。在解析完文件后，treeio 会将这 3 棵进化树存储到 multiPhylo 对象中。

```
r8s <- read.r8s(system.file("extdata/r8s", "H3_r8s_output.log",
package="treeio"))
r8s
```

```
## 3 phylogenetic trees
```

1.3.2.7　解析 RAxML 自举分析输出文件

RAxML 自举分析软件输出的是非标准的 Newick 文件，其在枝长后还添加了方括号来存储自举值。这导致此文件一般不能用常规的 Newick 解析器来解析，如 ape 包中的 read.tree() 函数等。而 treeio 中的 read.raxml() 函数则用于解析 RAxML 输出文件，利用该函数提取出自举值，存储为树的另一特征，方便后续进行可视化时将自举值显示于树上，或者根据自举值来为树的分支着色等。

```
raxml_file <- system.file("extdata/RAxML",
```

```
                               "RAxML_bipartitionsBranchLabels.H3",
                               package="treeio")
raxml <- read.raxml(raxml_file)
raxml
```

```
## 'treedata' S4 object that stored information
## of
##   '/home/ygc/R/library/treeio/extdata/RAxML/RAxML_
bipartitionsBranchLabels.H3'.
##
## ...@ phylo:
##
## Phylogenetic tree with 64 tips and 62 internal nodes.
##
## Tip labels:
##    A_Hokkaido_M1_2014_H3N2_2014,
## A_Czech_Republic_1_2014_H3N2_2014,
## FJ532080_A_California_09_2008_H3N2_2008,
## EU199359_A_Pennsylvania_05_2007_H3N2_2007,
## EU857080_A_Hong_Kong_CUHK69904_2006_H3N2_2006,
## EU857082_A_Hong_Kong_CUHK7047_2005_H3N2_2005, ...
##
## Unrooted; includes branch lengths.
##
## with the following features available:
##    'bootstrap'.
```

1.3.2.8 解析 NHX 树

NHX 是一种 Newick 的拓展格式，由于在 Newick 的基础上引进了 NHX 标签，因此 NHX 文件能存储更多信息。NHX 文件在系统发育学的软件中比较常见，如 PHYLDOG[14] 与 RevBayes[35] 就是使用 NHX 格式来存储统计推论的。而 treeio 中的 read.nhx() 函数则用于解析 NHX 格式的文件。下面的代码展示了如何使用 treeio 导入含有 PHYLDOG 推断出的关联数据的 NHX 树。

```
nhxfile <- system.file("extdata/NHX", "phyldog.nhx", package="treeio")
nhx <- read.nhx(nhxfile)
nhx
```

```
## 'treedata' S4 object that stored information
## of
```

```
##    '/home/ygc/R/library/treeio/extdata/NHX/phyldog.nhx'.
##
## ...@ phylo:
##
## Phylogenetic tree with 16 tips and 15 internal nodes.
##
## Tip labels:
##     Prayidae_D27SS7@2825365, Kephyes_ovata@2606431,
## Chuniphyes_multidentata@1277217,
## Apolemia_sp_@1353964, Bargmannia_amoena@263997,
## Bargmannia_elongata@946788, ...
##
## Rooted; includes branch lengths.
##
## with the following features available:
##     'Ev', 'ND', 'S'.
```

1.3.2.9 解析 Phylip 树

Phylip 格式内包含了 Phylip 序列格式的分类单元多序列比对信息，以及根据这些分类单元序列所建的 Newick 树。treeio 中的 read.phylip() 函数用于解析 Phylip 格式的文件。ggtree 中的 msaplot() 函数用于将多序列比对结果根据树的结构重新排列，并将其展示于树的右侧。除此之外，我们也可以结合 ggmsa 包来对多序列比对结果进行分析（另请参见本章参考文献 [36] 中的基本流程 5）。

```
phyfile <- system.file("extdata", "sample.phy", package="treeio")
phylip <- read.phylip(phyfile)
phylip
```

```
## 'treedata' S4 object that stored information
## of
##     '/home/ygc/R/library/treeio/extdata/sample.phy'.
##
## ...@ phylo:
##
## Phylogenetic tree with 15 tips and 13 internal nodes.
##
## Tip labels:
##     K, N, D, L, J, G, ...
##
## Unrooted; no branch lengths.
```

1.3.2.10 解析 EPA 和 pplacer 的输出文件

EPA[19] 与 pplacer[18] 输出的均为 Jplace 格式文件，而 treeio 中的 read.jplace() 函数则用于解析 Jplace 格式的文件。

```
jpf <- system.file("extdata/EPA.jplace", package="treeio")
jp <- read.jplace(jpf)
print(jp)
```

```
## 'treedata' S4 object that stored information
## of
##    '/home/ygc/R/library/treeio/extdata/EPA.jplace'.
##
## ...@ phylo:
##
## Phylogenetic tree with 493 tips and 492 internal nodes.
##
## Tip labels:
##    CIR000447A, CIR000479, CIR000078, CIR000083,
## CIR000070, CIR000060, ...
##
## Rooted; includes branch lengths.
##
## with the following features available:
##    'nplace'.
```

利用 treeio 计算每个分支上进化放置信息的数量，并将其存储于 nplace 特征中，以便在后续使用 ggtree[30] 进行可视化时用来设置分支线条的粗细或颜色。

1.3.2.11 解析 jtree 文件

jtree 格式是在 treeio 包 [29] 中定义的，是基于 JSON 格式的文件，用于支持树数据在不同编程语言之间的流通。我们可以使用 wirte.jtree() 函数来将进化树及其关联数据输出到同一个 jtree 文件中，所有支持 JSON 的解析器都可以解析 jtree 文件。其包含 3 个键（key）：tree、data 及 metadata。其中，tree 的值包含由 Newick 格式拓展而来的树文本，在枝长后还添加了花括号，用于存储边编号；data 的值包含了与分支或节点关联的数据；metadata 的值包含了额外的一些元信息。

```
jtree_file <- tempfile(fileext = '.jtree')
write.jtree(beast, file = jtree_file)
read.jtree(file = jtree_file)


## 'treedata' S4 object that stored information
## of
##    '/tmp/RtmpGClndg/file399657947c2f0.jtree'.
##
## ...@ phylo:
##
## Phylogenetic tree with 15 tips and 14 internal nodes.
##
## Tip labels:
##    K_2013, N_2010, D_1987, L_1980, J_1983, G_1992, ...
##
## Rooted; includes branch lengths.
##
## with the following features available:
##    'height', 'height_0.95_HPD', 'height_range',
## 'length', 'length_0.95_HPD', 'length_median',
## 'length_range', 'rate', 'rate_0.95_HPD',
## 'rate_median', 'rate_range', 'height_median',
## 'posterior'.
```

1.3.3 将其他树形对象转换为 phylo 对象或 treedata 对象

为了拓展 treeio[29]、tidytree 和 ggtree 的应用范围，treeio 为 as.phylo 和 as.treedata 定义了多种函数，以将类似 Phylo4d、pml 等树形对象转换为 phylo 对象或 treedata 对象。这样，用户就可以通过 treeio 包来更加便利地实现一些操作。例如，将相关数据映射至树结构上、将含有或不含有相关数据的树输出到同一个文件中、对带有数据的树进行可视化。这些转换函数让我们可以使用 tidytree 对整洁接口进行数据整理，或者使用 ggtree 利用图形语法（grammar of graphic）对树进行可视化。表 1.2 所示为将树形对象转换为 phylo 对象或 treedata 对象的函数。

表 1.2 将树形对象转换为 phylo 对象或 treedata 对象的函数

函数名称	支持对象	说明
as.phylo()	ggtree	将 ggtree 对象转换为 phylo 对象
	igraph	将 igraph object（仅支持树形图表）转换为 phylo 对象

续表

函数名称	支持对象	说明
as.phylo()	phylo4	将 phylo4 对象转换为 phylo 对象
	pvclust	将 pvclust 对象转换为 phylo 对象
	treedata	将 treedata 对象转换为 phylo 对象
as.treedata()	ggtree	将 ggtree 对象转换为 treedata 对象
	phylo4	将 phylo4 对象转换为 treedata 对象
	phylo4d	将 phylo4d 对象转换为 treedata 对象
	pml	将 pml 对象转换为 treedata 对象
	pvclust	将 pvclust 对象转换为 treedata 对象

这里使用 pml 对象来举例说明。pml 对象是在 phangorn 包中定义的，其中，pml() 函数用于根据给定的序列比对信息及模型计算出进化树的似然性，计算完成后可以使用 optm.pml() 函数对模型参数进行优化。最后输出一个 pml 对象，通过 treeio[29] 中的 as.treedata() 函数来将其转换为 treedata 对象，而 pml 对象中的氨基酸替换信息（pml 推断出的祖先序列）也会被存储到 treedata 对象中，以供后续分析中使用 ggtree 进行可视化。如图 1.4 所示，将 pml 对象转换为 treedata 对象后就可以使用 tidytree 来处理树及其相关数据，也可以使用 ggtree 和 ggtreeExtre 来对树及其关联数据进行可视化。

```
library(phangorn)
treefile <- system.file("extdata", "pa.nwk", package="treeio")
tre <- read.tree(treefile)
tipseqfile <- system.file("extdata", "pa.fas", package="treeio")
tipseq <- read.phyDat(tipseqfile,format="fasta")
fit <- pml(tre, tipseq, k=4)
fit <- optim.pml(fit, optNni=FALSE, optBf=T, optQ=T,
                optInv=T, optGamma=T, optEdge=TRUE,
                optRooted=FALSE, model = "GTR",
                control = pml.control(trace =0))

pmltree <- as.treedata(fit)
ggtree(pmltree) + geom_text(aes(x=branch, label=AA_subs, vjust=-.5))
```

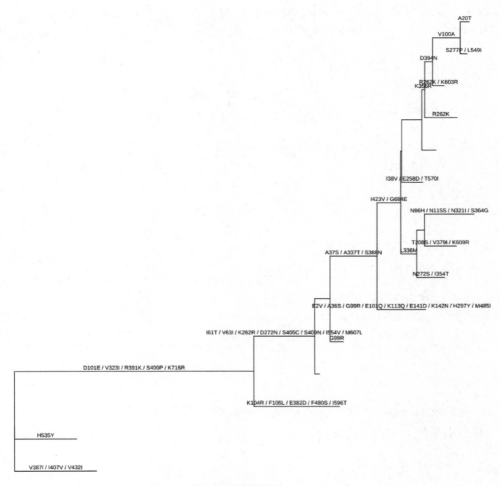

图 1.4 将 pml 对象转换为 treedata 对象

1.3.4 从 treedata 对象中获取信息

在使用 treeio 导入树后，可能会存在需要将存储于 treedata 对象中的信息提取出来的情况。因此，treeio 提供了多种存取器函数，并将 treedata 对象中存储的树结构，以及节点、分支相关的属性、特征等数据与它们对应的值提取出来。

首先，通过 treeio 中的 get.tree() 函数或 as.phylo() 函数将 treedata 对象转换为 phylo 对象。phylo 对象是 R 社区最基本的树对象，许多包都为 phylo 对象提供支持。

```r
beast_file <- system.file("examples/MCC_FluA_H3.tree", package="ggtree")
beast_tree <- read.beast(beast_file)

# 或者使用 get.tree() 函数

as.phylo(beast_tree)
```

```r
beast_file <- system.file("examples/MCC_FluA_H3.tree",
package="ggtree")
beast_tree <- read.beast(beast_file)
# 或者使用 get.tree() 函数
print(as.phylo(beast_tree), printlen=3)
```

```
## 
## Phylogenetic tree with 76 tips and 75 internal nodes.
## 
## Tip labels:
##   A/Hokkaido/30-1-a/2013, A/New_York/334/2004, A/New_York/463/2005, A/New_York/452/1999, A/New_York/238/2005, A/New_York/523/1998, ...
## 
## Rooted; includes branch lengths.
```

使用 get.fields() 函数可以将存储于 treedata 对象中的系统发育学相关属性或特征以向量的形式提取出来。

```r
get.fields(beast_tree)
```

```
##  [1] "height"           "height_0.95_HPD"
##  [3] "height_median"    "height_range"
##  [5] "length"           "length_0.95_HPD"
##  [7] "length_median"    "length_range"
##  [9] "posterior"        "rate"
## [11] "rate_0.95_HPD"    "rate_median"
## [13] "rate_range"
```

使用 get.data() 函数也可以提取这些信息，只是会以 tibble 的形式将数据提取出来。

```r
get.data(beast_tree)
```

```
## # A tibble: 151 x 14
##    height height_0.95_HPD height_median height_range
##     <dbl> <list>                  <dbl> <list>
##  1   19   <dbl [2]>                  19 <dbl [2]>
##  2   17   <dbl [2]>                  17 <dbl [2]>
##  3   14   <dbl [2]>                  14 <dbl [2]>
##  4   12   <dbl [2]>                  12 <dbl [2]>
##  5    9   <dbl [2]>                   9 <dbl [2]>
##  6   10   <dbl [2]>                  10 <dbl [2]>
##  7   10   <dbl [2]>                  10 <dbl [2]>
##  8   10.8 <dbl [2]>                10.8 <dbl [2]>
##  9    9   <dbl [2]>                   9 <dbl [2]>
## 10    9   <dbl [2]>                   9 <dbl [2]>
## # … with 141 more rows, and 10 more variables:
## #   length <dbl>, length_0.95_HPD <list>,
## #   length_median <dbl>, length_range <list>,
## #   posterior <dbl>, rate <dbl>, rate_0.95_HPD <list>,
## #   rate_median <dbl>, rate_range <list>, node <int>
```

如果用户只对使用 get.fields() 函数返回的特征或属性的子集感兴趣，则可以通过提取 get.data() 函数输出结果中的信息来实现，或者直接使用"["或"[["来提取数据的子集。

```
beast_tree[, c("node", "height")]
```

```
## # A tibble: 151 x 2
##     node height
##    <int>  <dbl>
##  1    10   19
##  2     9   17
##  3    36   14
##  4    31   12
##  5    29    9
##  6    28   10
##  7    39   10
##  8    90   10.8
##  9    16    9
## 10     2    9
## # … with 141 more rows
```

```
head(beast_tree[["height_median"]])
```

```
## height_median1 height_median2 height_median3
##            19              17              14
## height_median4 height_median5 height_median6
##            12               9              10
```

1.4 总结

目前，用于推断分子进化数据（如祖先状态、分子钟分析、选择压力等）的软件越来越多，但还是缺少一个统一的文件格式来存储这些不同系统发育学软件分析出的数据。大多数软件都是自己单独设计一个输出格式，而这些格式之间又互不兼容，导致解析不同软件的输出变得十分困难，使用多种软件一起联合分析的过程也变得异常烦琐。treeio 包 [29] 就是为了解决这个问题而生的。它提供了一系列函数来解析不同格式的系统发育学数据文件，也提供了一系列函数来将树形对象转换为 phylo 对象或 treedata 对象。这些系统发育学数据在整合后可以更好地进行接下来的数据探索与比较。目前来说，分子进化学领域的大多数软件都相对独立，不能很好地兼容其他软件的输入与输出，它们大多只负责自己的分析部分，而没有太过于注重输出的数据能否被其他软件读取。同时，缺少一个能将这些来自不同软件的输出结果整合到一起的工具。如果能将这些数据有效地整合起来，就能使我们对研究目标有一个更加全面的认识与理解，从而发现新的系统模式，或者提出新的假说。

在分子进化学背景下，利用进化树进行进化模式判别的应用范围越来越广泛，也有更多不同学科的学者开始将进化树应用于本学科的研究。例如，空间生态学家可以将研究生物的地理位置映射到进化树上，从而探究这些物种在生物地理学上的异同 [37]；流行病学家可以将病原体的采样时间与采样地点映射到进化树上，以在时间与空间层面上对疾病进行传播动力学研究 [38]；微生物学家可以在确定不同致病菌株的致病性后，将它们映射到进化树上，找出致病性的决定因素 [39]；基因组科学家可以使用进化树对宏基因组测序数据进行分类学层面的分类 [40]。对于他们来说，通过 treeio 这款功能强大的软件，能够将不同种类的数据导入 R 中，并与进化树关联起来，从而推动"发育动力学"的研究，也就是系统发育学相关的研究。通过 treeio 也可以将多种元数据（如时间、地理位置、基因型、流行病学信息等）与分析结果（如选择压力、进化速率等）整合起来，为学者提供了更全面研究生物学的工具。在流感领域的研究中，已经有人尝试将这些不同的元数据与分析结果映射到同一棵进化树及进化时间尺度[41]，来进行流感病毒的遗传动力学研究。

1.5 本章练习题

1. 简述系统发育树的作用及其构建方法。
2. 系统发育树的文件格式一般有哪些？并详述最为广泛使用的格式。
3. 简述哪些函数用于读取以下输出文件。

（1）Newick 输出文件。

（2）RAxML 输出文件。

（3）BEAST 输出文件。

（4）MrBayes 输出文件。

（5）HyPhy 输出文件。

（6）CODEML 输出文件。

（7）mega 输出文件。

（8）r8s 输出文件。

参考文献

[1] Fitch W M. Toward defining the course of evolution: minimum change for a specific tree topology[J]. Systematic Zoology, 1971, 20(4): 406-416.

[2] Felsenstein J. Evolutionary trees from DNA sequences: a maximum likelihood approach[J]. J Mol Evol, 1981, 17(6): 368-376.

[3] Rannala B Y Z. Probability distribution of molecular evolutionary trees: a new method of phylogenetic inference[J]. Journal of Molecular Evolution, 1996, 43(3): 304-311.

[4] Felsenstein J. Cases in which parsimony or compatibility methods will be positively misleading. [J]. Journal of Molecular Evolution, 1978, 27(4): 401-410.

[5] Arenas M. Trends in substitution models of molecular evolution[J]. Front Genet, 2015,6:319.
[6] Yang Z. Maximum likelihood phylogenetic estimation from DNA sequences with variable rates over sites: approximate methods[J]. J Mol Evol, 1994, 39(3): 306-314.
[7] Shoemaker J S, Fitch W M. Evidence from nuclear sequences that invariable sites should be considered when sequence divergence is calculated[J]. Mol Biol Evol, 1989, 6(3): 270-289.
[8] Lemmon A R, Moriarty E C. The importance of proper model assumption in bayesian phylogenetics[J]. Syst Biol, 2004, 53(2): 265-277.
[9] Maddison D R, Swofford D L, Maddison W P. NEXUS: an extensible file format for systematic information[J]. Syst Biol, 1997, 46(4): 590-621.
[10] Felsenstein J. PHYLIP - Phylogeny Inference Package (Version 3.2)[J]. Cladistics, 1989, 5: 164-166.
[11] Retief J D. Phylogenetic analysis using PHYLIP[J]. Methods Mol Biol, 2000,132:243-258.
[12] Wilgenbusch J C, Swofford D. Inferring evolutionary trees with PAUP*[J]. Curr Protoc Bioinformatics, 2003(1): 6.4.1-6.4.28.
[13] Schmidt H A, Strimmer K, Vingron M, et al. TREE-PUZZLE: maximum likelihood phylogenetic analysis using quartets and parallel computing[J]. Bioinformatics, 2002,18(3):502-504.
[14] Boussau B, Szollosi G J, Duret L, et al. Genome-scale coestimation of species and gene trees[J]. Genome Res, 2013,23(2):323-330.
[15] Höhna S, Landis M J, Heath T A, et al. RevBayes: Bayesian phylogenetic inference using graphical models and an interactive model-specification language[J]. Systematic Biology, 2016, 65(4): 726-736.
[16] Zmasek C M, Eddy S R. ATV: display and manipulation of annotated phylogenetic trees[J]. Bioinformatics, 2001, 17(4): 383-384.
[17] Matsen F A, Hoffman N G, Gallagher A, et al. A format for phylogenetic placements[J]. PLoS One, 2012, 7(2): e31009.
[18] Matsen F A, Kodner R B, Armbrust E V. pplacer: linear time maximum-likelihood and Bayesian phylogenetic placement of sequences onto a fixed reference tree[J]. BMC Bioinformatics, 2010, 11: 538.
[19] Berger S A, Krompass D, Stamatakis A. Performance, accuracy, and web server for evolutionary placement of short sequence reads under maximum likelihood[J]. Syst Biol, 2011, 60(3): 291-302.
[20] Stamatakis A. RAxML version 8: a tool for phylogenetic analysis and post-analysis of large phylogenies[J]. Bioinformatics, 2014, 30(9): 1312-1313.
[21] Bouckaert R, Heled J, Kuhnert D, et al. BEAST 2: a software platform for Bayesian evolutionary analysis[J]. PLoS Comput Biol, 2014, 10(4): e1003537.
[22] Kumar S, Stecher G, Tamura K. MEGA7: molecular evolutionary genetics analysis version 7.0 for bigger datasets[J]. Mol Biol Evol, 2016, 33(7): 1870-1874.
[23] Huelsenbeck J P, Ronquist F. MRBAYES: Bayesian inference of phylogenetic trees[J]. Bioinformatics, 2001, 17(8): 754-755.

[24] Pond S L, Frost S D, Muse S V. HyPhy: hypothesis testing using phylogenies[J]. Bioinformatics, 2005, 21(5): 676-679.

[25] Sanderson M J. r8s: inferring absolute rates of molecular evolution and divergence times in the absence of a molecular clock[J]. Bioinformatics, 2003, 19(2): 301-302.

[26] Yang Z. PAML 4: phylogenetic analysis by maximum likelihood[J]. Mol Biol Evol, 2007, 24(8): 1586-1591.

[27] Goldman N, Yang Z. A codon-based model of nucleotide substitution for protein-coding DNA sequences[J]. Mol Biol Evol, 1994, 11(5): 725-736.

[28] Team R C. R: A language and environment for statistical computing. R Foundation for Statistical Computing[EB/OL]. https://www.R-project.org/.

[29] Wang L, Lam T T, Xu S, et al. treeio: an R package for phylogenetic tree input and output with richly annotated and associated data[J]. Molecular Biology and Evolution, 2020,37(2):599-603.

[30] Yu G, Smith D K, Zhu H, et al. ggtree: an R package for visualization and annotation of phylogenetic trees with their covariates and other associated data[J]. Methods in Ecology and Evolution, 2016, 8(1): 28-36.

[31] Xu S, Dai Z, Guo P, et al. ggtreeExtra: compact visualization of richly annotated phylogenetic data[J]. Mol Biol Evol, 2021, 38(9): 4039-4042.

[32] Paradis E, Claude J, Strimmer K. APE: analyses of phylogenetics and evolution in R language[J]. Bioinformatics, 2004, 20(2): 289-290.

[33] Schliep K P. phangorn: phylogenetic analysis in R[J]. Bioinformatics, 2011, 27(4): 592-593.

[34] Yu G, Lam T T, Zhu H, et al. Two methods for mapping and visualizing associated data on phylogeny using ggtree[J]. Mol Biol Evol, 2018, 35(12): 3041-3043.

[35] Hohna S, Heath T A, Boussau B, et al. Probabilistic graphical model representation in phylogenetics[J]. Syst Biol, 2014, 63(5): 753-771.

[36] Yu G. Using ggtree to visualize data on tree-like structures[J]. Current protocols in bioinformatics, 2020, 69(1): e96.

[37] Schön I S R M. Age and origin of Australian Bennelongia (Crustacea, Ostracoda)[J]. Hydrobiologia, 2015, 750(1): 125-146.

[38] He Y Q, Chen L, Xu W B, et al. Emergence, circulation, and spatiotemporal phylogenetic analysis of coxsackievirus a6- and coxsackievirus a10-associated hand, foot, and mouth disease infections from 2008 to 2012 in Shenzhen, China[J]. J Clin Microbiol, 2013, 51(11): 3560-3566.

[39] Bosi E, Monk J M, Aziz R K, et al. Comparative genome-scale modelling of Staphylococcus aureus strains identifies strain-specific metabolic capabilities linked to pathogenicity[J]. Proc Natl Acad Sci U S A, 2016, 113(26): E3801-E3809.

[40] Gupta A, Sharma V K. Using the taxon-specific genes for the taxonomic classification of bacterial genomes[J]. BMC Genomics, 2015, 16: 396.

[41] Lam T T, Zhou B, Wang J, et al. Dissemination, divergence and establishment of H7N9 influenza viruses in China[J]. Nature, 2015, 522(7554): 102-105.

第 2 章　操作含有关联数据的树

2.1　使用 tidy 接口操作树数据

通过 tidytree 包，我们可以把通过 treeio[1] 解析或整合的所有树数据都转换为整洁数据框（tidy data frame）。tidytree 包还提供了整洁接口（tidy interface），也用于操作含有关联数据的树。例如，我们不仅可以将外部数据关联到进化树，还能将不同来源的进化数据使用 tidyverse 函数进行合并。经过处理的树数据还可以被重新转换为 treedata 对象，并输出至单个树文件中，也可以在 R 中继续进行进一步的分析，或者使用 ggtree[2] 和 ggtreeExtra[3] 对其进行可视化。

2.1.1　phylo 对象

在 ape 包[4]中定义的 phylo 类是在 R 中进行系统发育分析的基础。系统发育分析中用到的大多数 R 包都依赖于 phylo 对象。tidytree 包中的 as_tibble() 函数用于将 phylo 对象转换为一个整洁数据框，即一个 tbl_tree 对象。

```
library(ape)
set.seed(2017)
tree <- rtree(4)
tree
```

```
##
## Phylogenetic tree with 4 tips and 3 internal nodes.
##
## Tip labels:
##   t4, t1, t3, t2
##
## Rooted; includes branch lengths.
```

```
x <- as_tibble(tree)
x
```

```
## # A tibble: 7 × 4
##   parent  node branch.length label
##    <int> <int>         <dbl> <chr>
## 1      5     1       0.435   t4
## 2      7     2       0.674   t1
## 3      7     3       0.00202 t3
## 4      6     4       0.0251  t2
## 5      5     5      NA       <NA>
## 6      5     6       0.472   <NA>
## 7      6     7       0.274   <NA>
```

通过 as.phylo() 函数，我们还可以将 tbl_tree 对象转换为 phylo 对象。

```
as.phylo(x)
```

```
##
## Phylogenetic tree with 4 tips and 3 internal nodes.
##
## Tip labels:
##   t4, t1, t3, t2
##
## Rooted; includes branch lengths.
```

我们可以通过 tbl_tree 对象简单、高效地处理树及关联数据。例如，我们可以使用 dplyr 中的 full_join() 函数将进化性状信息与系统发育树相关联。

```
d <- tibble(label = paste0('t', 1:4),
            trait = rnorm(4))

y <- full_join(x, d, by = 'label')
y
```

```
## # A tibble: 7 × 5
##   parent  node branch.length label trait
##    <int> <int>         <dbl> <chr> <dbl>
## 1      5     1         0.435 t4    0.943
```

```
## 2     7    2    0.674    t1    -0.171
## 3     7    3    0.00202  t3     0.570
## 4     6    4    0.0251   t2    -0.283
## 5     5    5    NA       <NA>   NA
## 6     5    6    0.472    <NA>   NA
## 7     6    7    0.274    <NA>   NA
```

2.1.2 treedata 对象

tidytree 包定义了 treedata 类用于存储进化树及其关联数据。将外部数据映射到树结构后，tbl_tree 对象可以被转换为 treedata 对象。

```
as.treedata(y)
```

```
## 'treedata' S4 object'.
## 
## ...@ phylo:
## 
## Phylogenetic tree with 4 tips and 3 internal nodes.
## 
## Tip labels:
##   t4, t1, t3, t2
## 
## Rooted; includes branch lengths.
## 
## with the following features available:
##   'trait'.
```

treeio 包[1]中的 treedata 类用于存储常用软件（如 BEAST、EPA、HyPhy、MrBayes、PAML、PHYLDOG、PPLACER、r8s、RAxML 和 RevBayes）推断的演化证据。

tidytree 包中的 as_tibble() 函数用于将 treedata 对象转换为简洁数据框。进化树的结构和进化推论都会被存储到 tbl_tree 对象中，使得我们能更便捷地处理不同软件推断的进化统计数据及将外部数据关联到相同的树结构中。

```
y %>% as.treedata %>% as_tibble
```

```
## # A tibble: 7 × 5
##   parent  node branch.length label  trait
```

```
##     <int> <int>       <dbl> <chr>   <dbl>
## 1      5     1       0.435  t4      0.943
## 2      7     2       0.674  t1     -0.171
## 3      7     3       0.00202 t3     0.570
## 4      6     4       0.0251 t2     -0.283
## 5      5     5       NA     <NA>    NA
## 6      5     6       0.472  <NA>    NA
## 7      6     7       0.274  <NA>    NA
```

2.1.3 访问相关节点

我们可以使用 dplyr 包中的函数来操作树数据。此外，tidytree 还提供了 child()、parent()、offspring()、ancestor()、sibling() 和 MRCA() 几个函数来筛选相关的节点。

这些函数需要接收一个 tbl_tree 对象及一个选定的节点。我们可以通过节点编号或节点标签来对该节点进行选择。

```
child(y, 5)
```

```
## # A tibble: 2 × 5
##   parent  node branch.length label trait
##    <int> <int>         <dbl> <chr> <dbl>
## 1      5     1         0.435 t4    0.943
## 2      5     6         0.472 <NA>  NA
```

```
parent(y, 2)
```

```
## # A tibble: 1 × 5
##   parent  node branch.length label trait
##    <int> <int>         <dbl> <chr> <dbl>
## 1      6     7         0.274 <NA>  NA
```

```
offspring(y, 5)
```

```
## # A tibble: 6 × 5
##   parent  node branch.length label trait
##    <int> <int>         <dbl> <chr> <dbl>
## 1      5     1         0.435 t4    0.943
```

```
## 2        7     2          0.674   t1     -0.171
## 3        7     3          0.00202 t3      0.570
## 4        6     4          0.0251  t2     -0.283
## 5        5     6          0.472   <NA>    NA
## 6        6     7          0.274   <NA>    NA
```

```
ancestor(y, 2)
```

```
## # A tibble: 3 × 5
##   parent  node branch.length label trait
##    <int> <int>         <dbl> <chr> <dbl>
## 1      5     5         NA    <NA>   NA
## 2      5     6          0.472 <NA>  NA
## 3      6     7          0.274 <NA>  NA
```

```
sibling(y, 2)
```

```
## # A tibble: 1 × 5
##   parent  node branch.length label trait
##    <int> <int>         <dbl> <chr> <dbl>
## 1      7     3       0.00202 t3    0.570
```

```
MRCA(y, 2, 3)
```

```
## # A tibble: 1 × 5
##   parent  node branch.length label trait
##    <int> <int>         <dbl> <chr> <dbl>
## 1      6     7         0.274 <NA>   NA
```

上述所有函数在 treeio 中也都有所实现，可用于操作 phylo 对象与 treedata 对象。用户可以尝试使用这些函数来访问树对象的相关节点。例如，使用以下命令输出所选的 5 号节点的子节点。

```
child(tree, 5)
```

```
## [1] 1 6
```

需要注意的是，在使用处理树对象的函数时会输出相关的节点编号，而在使用处理 tbl_tree 对象的函数时则会输出包含相关信息的 tibble 对象。

2.2 数据整合

2.2.1 整合树数据

treeio 包[1] 作为 R 中进行系统发育分析的基础，能够将不同的分析程序推断的各类系统发育数据导入 R 并在 R 中使用。例如，由 CODEML 推断的祖先序列与 dN/dS 值，以及由 BEAST 或 MrBayes 推断的进化枝支持度（后验）的值。treeio 包还支持将外部数据关联到系统发育树。它能将外部的系统发育数据带入 R 社区，以便在 R 中进行更进一步的分析。此外，treeio 包可以将多棵系统发育树整合为一棵保留了各个分支与节点属性的进化树。正因如此，我们可以将某个属性（如替换速率）映射到同一个节点和分支的另一个属性上（如 dN/dS），以便于进行比对及进一步的计算 [2, 5]。

在此示例中，我们使用了一个已被发表的数据集，其中包含了人猪甲型流感病毒谱系中的 76 条 H3 血凝素基因序列 [6]，在此用于演示如何比较不同软件推断的进化统计数据。数据集先经由 BEAST 进行重新分析，并进行时间尺度分析，再使用 CODEML 估算同义及非同义替换。

我们首先分别使用 read.beast() 函数解析 BEAST 的输出文件，使用 read.codeml() 函数解析 CODEML 的输出文件，这两个函数分别输出两个 treedata 对象——beast_tree 对象和 codeml_tree 对象。然后使用 merge_tree() 函数合并两个包含节点或分支关联数据集的 treedata 对象——beast_tree 对象和 codeml_tree 对象。

```
beast_file <- system.file("examples/MCC_FluA_H3.tree",
package="ggtree")
rst_file <- system.file("examples/rst", package="ggtree")
mlc_file <- system.file("examples/mlc", package="ggtree")
beast_tree <- read.beast(beast_file)
codeml_tree <- read.codeml(rst_file, mlc_file)

merged_tree <- merge_tree(beast_tree, codeml_tree)
merged_tree
```

```
## 'treedata' S4 object that stored information
## of
##    '/home/ygc/R/library/ggtree/examples/MCC_FluA_H3.tree',
##    '/home/ygc/R/library/ggtree/examples/rst',
##    '/home/ygc/R/library/ggtree/examples/mlc'.
##
## ...@ phylo:
##
## Phylogenetic tree with 76 tips and 75 internal nodes.
##
## Tip labels:
##    A/Hokkaido/30-1-a/2013, A/New_York/334/2004,
## A/New_York/463/2005, A/New_York/452/1999,
## A/New_York/238/2005, A/New_York/523/1998, ...
##
## Rooted; includes branch lengths.
##
## with the following features available:
##    'height', 'height_0.95_HPD', 'height_median',
## 'height_range', 'length', 'length_0.95_HPD',
## 'length_median', 'length_range', 'posterior', 'rate',
## 'rate_0.95_HPD', 'rate_median', 'rate_range', 'subs',
## 'AA_subs', 't', 'N', 'S', 'dN_vs_dS', 'dN', 'dS',
## 'N_x_dN', 'S_x_dS'.
```

在合并 beast_tree 对象和 codeml_tree 对象后，所有从 BEAST 和 CODEML 输出文件导入的节点或分支关联数据都会被存储在 merged_tree 对象中，且可以用来做各种后续分析。接下来，我们使用 tidytree 包将此树对象转换为简洁数据框，并以六边形散点图的形式对 CODEML 推断的 dN/dS、dN、dS，与在同一个分支上由 BEAST 推断的 rate（替换速率，以每年每个位点的替换数为单位）进行比较。

```
library(dplyr)
df <- merged_tree %>%
  as_tibble() %>%
  select(dN_vs_dS, dN, dS, rate) %>%
  subset(dN_vs_dS >=0 & dN_vs_dS <= 1.5) %>%
  tidyr::gather(type, value, dN_vs_dS:dS)
df$type[df$type == 'dN_vs_dS'] <- 'dN/dS'
df$type <- factor(df$type, levels=c("dN/dS", "dN", "dS"))
```

```
ggplot(df, aes(rate, value)) + geom_hex() +
  facet_wrap(~type, scale='free_y')
```

输出结果如图 2.1 所示。在合并 BEAST 和 CODEML 的输出结果后，来自两个分析程序的分支相关统计推断（如替换速率，dN/dS、dN、dS）将会基于同一个分支进行对比。dN/dS、dN 和 dS 与替换速率的相关性将会以六边形散点图的形式呈现。

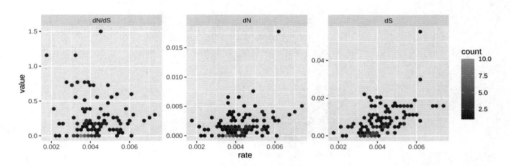

图 2.1　dN/dS、dN 和 dS 与替换速率的相关性

在此之后，我们就可以对这些分支或节点相关数据进行皮尔逊相关性（Pearson Correlation）检验。在本示例中，我们可以看到 dN 和 dS 与替换速率显著相关（由 p 值确定），而 dN/dS 和替换速率并没有显著相关性。

使用 merge_tree() 函数，我们不仅可以比对来自不同软件包基于同一个模型得出的分析结果，也可以比对来自不同或相同软件基于不同模型产生的分析结果。同时，merge_tree() 函数不仅用于支持不同软件包分析结果的整合，还用于将软件分析结果与外部数据关联起来。merge_tree() 函数具有管道连接性，可以将多个树对象依次合并为一个整体。

```
phylo <- as.phylo(beast_tree)
N <- Nnode2(phylo)
d <- tibble(node = 1:N, fake_trait = rnorm(N), another_trait = runif(N))
fake_tree <- treedata(phylo = phylo, data = d)
triple_tree <- merge_tree(merged_tree, fake_tree)
triple_tree


## 'treedata' S4 object that stored information
## of
```

```
##   '/home/ygc/R/library/ggtree/examples/MCC_FluA_H3.tree',
##   '/home/ygc/R/library/ggtree/examples/rst',
##   '/home/ygc/R/library/ggtree/examples/mlc'.
## 
## ...@ phylo:
## 
## Phylogenetic tree with 76 tips and 75 internal nodes.
## 
## Tip labels:
##   A/Hokkaido/30-1-a/2013, A/New_York/334/2004,
## A/New_York/463/2005, A/New_York/452/1999,
## A/New_York/238/2005, A/New_York/523/1998, ...
## 
## Rooted; includes branch lengths.
## 
## with the following features available:
##   'height', 'height_0.95_HPD', 'height_median',
## 'height_range', 'length', 'length_0.95_HPD',
## 'length_median', 'length_range', 'posterior', 'rate',
## 'rate_0.95_HPD', 'rate_median', 'rate_range', 'subs',
## 'AA_subs', 't', 'N', 'S', 'dN_vs_dS', 'dN', 'dS',
## 'N_x_dN', 'S_x_dS', 'fake_trait', 'another_trait'.
```

如上代码所示的 triple_tree 对象包含了从 BEAST 和 CODEML 获得的分析结果及外部来源的进化性状信息。所有这些信息都可用 ggtree[2] 和 ggtreeExtra[3] 来对树进行可视化注释。

2.2.2 将外部数据关联到系统发育树

除了上述提到的这些与进化树相关的分析结果，还存在着大量的异构数据（如表型数据、实验数据和临床数据等）需要被整合并关联到进化树上。例如，在对病毒演化的研究中，树中的节点可能会与流行病学信息（如地点、年份和亚型等）存在关联。在比较基因组学的研究中，我们也需要将功能注释信息映射到基因树上。为了便于数据整合，treeio 提供了 full_join() 函数来将外部数据关联到存储于 phylo 对象或 treedata 对象中的系统发育树上。需要注意的是，如果将外部数据关联到 phylo 对象，则会生成一个 treedata 对象来存储输入的 phylo 对象及其相关联的数据。除此之外，full_join() 函数还支持在简洁数据框（前文中提到的

tbl_tree 对象）水平及 ggtree 水平使用（参见第 7 章或本章参考文献 [5]）。

下面示例在计算自举值后，通过匹配节点编号将其与树（一个 phylo 对象）关联起来。

```
library(ape)
data(woodmouse)
d <- dist.dna(woodmouse)
tr <- nj(d)
bp <- boot.phylo(tr, woodmouse, function(x) nj(dist.dna(x)))
```

```
## Running bootstraps:           100/100
## Calculating bootstrap values... done.
```

```
bp2 <- tibble(node=1:Nnode(tr) + Ntip(tr), bootstrap = bp)
full_join(tr, bp2, by="node")
```

```
## 'treedata' S4 object'.
##
## ...@ phylo:
##
## Phylogenetic tree with 15 tips and 13 internal nodes.
##
## Tip labels:
##    No305, No304, No306, No0906S, No0908S, No0909S, ...
##
## Unrooted; includes branch lengths.
##
## with the following features available:
##    'bootstrap'.
```

接下来的示例通过匹配叶节点标签将进化性状与树（一个 treedata 对象）合并。

```
file <- system.file("extdata/BEAST", "beast_mcc.tree",
package="treeio")
beast <- read.beast(file)
x <- tibble(label = as.phylo(beast)$tip.label, trait =
rnorm(Ntip(beast)))
full_join(beast, x, by="label")
```

```
## 'treedata' S4 object that stored information
## of
##   '/home/ygc/R/library/treeio/extdata/BEAST/beast_mcc.tree'.
##
## ...@ phylo:
##
## Phylogenetic tree with 15 tips and 14 internal nodes
##
## Tip labels:
##   A_1995, B_1996, C_1995, D_1987, E_1996, F_1997, ...
##
## Rooted; includes branch lengths.
##
## with the following features available:
##   'height', 'height_0.95_HPD', 'height_median',
## 'height_range', 'length', 'length_0.95_HPD',
## 'length_median', 'length_range', 'posterior', 'rate',
## 'rate_0.95_HPD', 'rate_median', 'rate_range', 'trait'.
```

由于处理 phylo 对象的函数过于零碎，操作树对象往往会显得非常困难，更不用说将外部数据关联到系统发育树的结构。而使用 treeio 包[1]可以很轻松地将不同来源的树数据整合到一起。tidytree 包也可以通过整洁数据原则（tidy data principles）更为便捷地对树进行操作，并且能与已经被广泛使用的工具进行兼容，如 dplyr、tidyr、ggplot2 和 ggtree[2] 等。

2.2.3 对分类单元进行分组

groupClade() 函数和 groupOTU() 函数可用于将分类单元分组信息添加到输入的树对象中。tidytree、treeio 和 ggtree 均实现了这两个函数，并分别支持为 tbl_tree 对象、phylo 对象、treedata 对象和 ggtree 对象添加分组信息。这些分组信息可以直接用于使用 ggtree 进行树的可视化（如根据分组信息为树着色）。

2.2.3.1 groupClade() 函数

groupClade() 函数用于接收一个内部节点或一个由多个内部节点组成的向量作为输入，来为选定的一个或多个进化枝添加分组信息。

```
nwk <- '(((((((A:4,B:4):6,C:5):8,D:6):3,E:21):10,((F:4,G:12):14
```

```
,H:8):13):
        13,((I:5,J:2):30,(K:11,L:11):2):17):4,M:56);'
tree <- read.tree(text=nwk)

groupClade(as_tibble(tree), c(17, 21))
```

```
## # A tibble: 25 × 5
##    parent  node branch.length label group
##     <int> <int>         <dbl> <chr> <fct>
## 1      20     1             4 A     1
## 2      20     2             4 B     1
## 3      19     3             5 C     1
## 4      18     4             6 D     1
## 5      17     5            21 E     1
## 6      22     6             4 F     2
## 7      22     7            12 G     2
## 8      21     8             8 H     2
## 9      24     9             5 I     0
## 10     24    10             2 J     0
## # ⋯ with 15 more rows
```

2.2.3.2　groupOTU() 函数

下面介绍 groupOTU() 函数示例。

```
set.seed(2017)
tr <- rtree(4)
x <- as_tibble(tr)
## 可以输入节点编号或节点标签
groupOTU(x, c('t1', 't4'), group_name = "fake_group")
```

```
## # A tibble: 7 × 5
##    parent  node branch.length label fake_group
##     <int> <int>         <dbl> <chr>      <fct>
## 1       5     1         0.435 t4             1
## 2       7     2         0.674 t1             1
## 3       7     3         0.00202 t3           0
## 4       6     4         0.0251 t2            0
## 5       5     5         NA    <NA>           1
## 6       5     6         0.472 <NA>           1
## 7       6     7         0.274 <NA>           1
```

groupClade() 函数和 groupOTU() 函数都适用于 tbl_tree 对象、phylo 对象、treedata 对象与 ggtree 对象。下面是对 phylo 树对象使用 groupOTU() 函数的示例。

```
groupOTU(tr, c('t2', 't4'), group_name = "fake_group") %>%
  as_tibble
```

```
## # A tibble: 7 × 5
##   parent  node branch.length label fake_group label fake_group
##    <int> <int>         <dbl> <chr> <fct>
## 1      5     1         0.435 t4    1
## 2      7     2         0.674 t1    0
## 3      7     3         0.00202 t3  0
## 4      6     4         0.0251 t2   1
## 5      5     5        NA     <NA>  1
## 6      5     6         0.472 <NA>  1
## 7      6     7         0.274 <NA>  0
```

groupOTU() 函数用于从输入的节点追溯至它们最近的共同祖先。在此示例中，节点 1、节点 4、节点 5 和节点 6 被分配到同一个组中（4(t2)->6->5 和 1(t4)->5）。

处于不同进化枝内的相关分类操作分类单元（Operational Taxonomic Units，OTUs）也能被分配到同一个组中。这些 OTUs 可以是单系群（在同一进化枝内）、多系群或旁系群。

```
cls <- list(c1=c("A", "B", "C", "D", "E"),
            c2=c("F", "G", "H"),
            c3=c("L", "K", "I", "J"),
            c4="M")
as_tibble(tree) %>% groupOTU(cls)
```

```
## # A tibble: 25 × 5
##   parent  node branch.length label group label fake_group
##    <int> <int>         <dbl> <chr> <fct>
## 1     20     1             4 A     c1
## 2     20     2             4 B     c1
## 3     19     3             5 C     c1
## 4     18     4             6 D     c1
## 5     17     5            21 E     c1
```

```
## 6        22        6              4 F        c2
## 7        22        7             12 G        c2
## 8        21        8              8 H        c2
## 9        24        9              5 I        c3
## 10       24       10              2 J        c3
## # ... with 15 more rows
```

如果在追溯到最近的共同祖先时出现冲突，则用户可以将 overlap 参数设置为 origin（以第一个追溯到的祖先为准）、overwrite（默认值，以最后一个追溯到的祖先为准）或 abandon（取消对冲突 OTUs 的分组）[①]。

2.3 重新设定树的根节点

系统发育树可以根据指定的 outgroup（外群）参数重新设定根节点。ape 包提供了 root() 函数来为存储在 phylo 对象中的树重新设定根节点。同样地，treeio 包也为 treedata 对象提供了 root() 函数。该函数基于 ape 包中实现的 root() 函数，支持通过特定的 outgroup 或 node 来重新设定含有相关数据的进化树的根节点。

在下面示例中，首先使用 left_join() 函数将外部数据关联到树，并将所有信息存储到 treedata 对象 trda 中。

```
library(ggtree)
library(treeio)
library(tidytree)
library(TDbook)
# 从 TDbook 中加载 tree_boots, df_tip_data, 以及 df_inode_data 数据
trda <- tree_boots %>%
        left_join(df_tip_data, by=c("label" = "Newick_label")) %>%
        left_join(df_inode_data, by=c("label" = "newick_label"))
trda

## 'treedata' S4 object'.
##
## ...@ phylo:
##
## Phylogenetic tree with 7 tips and 6 internal nodes.
##
```

---
① 相关讨论请参见"外链资源"文档中第 2 章第 2 条

```
## Tip labels:
## Rangifer_tarandus, Cervus_elaphus, Bos_taurus,
## Ovis_orientalis, Suricata_suricatta,
## Cystophora_cristata, ...
## Node labels:
##   Mammalia, Artiodactyla, Cervidae, Bovidae,
## Carnivora, Caniformia
##
## Rooted; includes branch lengths.
##
## with the following features available:
##   '', 'vernacularName', 'imageURL', 'imageLicense',
## 'imageAuthor', 'infoURL', 'mass_in_kg',
## 'trophic_habit', 'ncbi_taxid', 'rank',
## 'vernacularName.y', 'infoURL.y', 'rank.y',
## 'bootstrap', 'posterior'.
```

然后重新设定含有相关数据的树的根节点，并将其关联数据正确地映射到对应的分支和节点。如图 2.2 所示，原始树（A）和重新设定根节点后的树（B）在重新设定根节点时也会将其相关数据正确映射到树的分支或节点。A 和 B 分别呈现了树在将根节点改为"Suricata_suricatta"的父节点前后的样子。该图是使用 **ggtree** 进行可视化的。

```
# 重新设定树的根节点
trda2 <- root(trda, outgroup = "Suricata_suricatta", edgelabel = TRUE)
# 原始树
p1 <- trda %>%
    ggtree() +
    geom_nodelab(
      mapping = aes(
        x = branch,
        label = bootstrap
      ),
      nudge_y = 0.36
    ) +
    xlim(-.1, 4) +
    geom_tippoint(
      mapping = aes(
        shape = trophic_habit,
        color = trophic_habit,
```

```r
      size = mass_in_kg
    )
  ) +
  scale_size_continuous(range = c(3, 10)) +
  geom_tiplab(
    offset = .14,
  ) +
  geom_nodelab(
    mapping = aes(
      label = vernacularName.y,
      fill = posterior
    ),
    geom = "label"
  ) +
  scale_fill_gradientn(colors = RColorBrewer::brewer.pal(3, "YlGnBu")) +
  theme(legend.position = "right")
# 重新设定根节点后的树
p2 <- trda2 %>%
  ggtree() +
  geom_nodelab(
    mapping = aes(
      x = branch,
      label = bootstrap
    ),
    nudge_y = 0.36
  ) +
  xlim(-.1, 5) +
  geom_tippoint(
    mapping = aes(
      shape = trophic_habit,
      color = trophic_habit,
      size = mass_in_kg
    )
  ) +
  scale_size_continuous(range = c(3, 10)) +
  geom_tiplab(
    offset = .14,
  ) +
  geom_nodelab(
    mapping = aes(
      label = vernacularName.y,
```

```
        fill = posterior
    ),
    geom = "label"
) +
    scale_fill_gradientn(colors = RColorBrewer::brewer.pal(3,
"YlGnBu")) +
    theme(legend.position = "right")
plot_list(p1, p2, tag_levels='A', ncol=2)
```

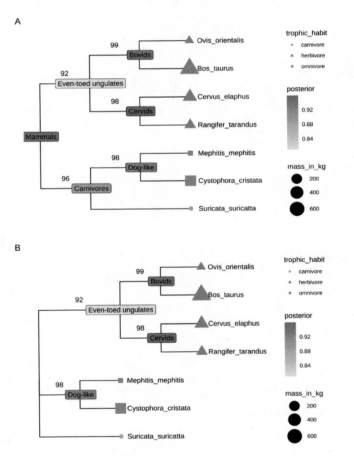

图 2.2　重新设定含有关联数据的树的根节点

outgroup 参数表示指定的新的外群，可以是节点标签（字符）或节点编号。如果只有一个值，则以对应叶节点的父节点作为新的根节点。如果有多个值，则使用它们最近共同祖先的节点作为新的根节点。需要注意的是，如果遇到需要将节点标签视作为分支标签的情况,则需要将 edgelabel 参数的值设置为 TRUE,以便正确地返回节点与关联数据之间的关系。有关重新设置根节点的更多内容请参考本章参考文献 [7]。

2.4 重新调整分支标尺

我们可以将系统发育数据与其他数据合并,以便用于联合分析(见图 2.3),也可以利用它们对树进行可视化,在树的结构上呈现复杂的注释,以便于更直观地揭示它们的演化模式。存储在 treedata 对象中的所有数值数据都可用于重新调整分支标尺。例如,CODEML 推断的 dN/dS、dN 及 dS,这些统计数据都可以用作枝长。图 2.3A 是以时间作为标尺的树(从根节点开始计时),图 2.3B 与图 2.3C 分别是以 dN 作为枝长调整标尺,以及以替换速率作为枝长调整标尺。除此之外,它们还可以用来给树着色,或者投影到二维空间以创建二维树或表型图。

```
p1 <- ggtree(merged_tree) + theme_tree2()
p2 <- ggtree(rescale_tree(merged_tree, 'dN')) + theme_tree2()
p3 <- ggtree(rescale_tree(merged_tree, 'rate')) + theme_tree2()

plot_list(p1, p2, p3, ncol=3, tag_levels='A')
```

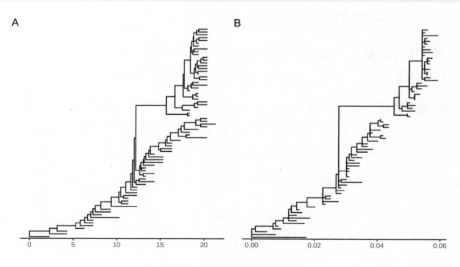

图 2.3 重新设定分支标尺

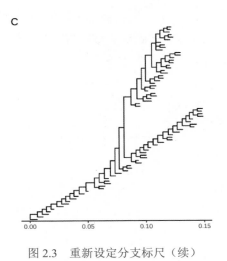

图 2.3　重新设定分支标尺（续）

除了可以使用 rescale_tree() 函数修改树对象中的分支长度，用户还可以直接在 ggtree() 函数中指定一个数值变量作为分支长度。

2.5　对包含数据的树取子集

2.5.1　删除系统发育树中的叶节点

有时我们需要从进化树中删除选定的叶节点，其主要原因是：序列本身质量低、序列组装出错、部分序列比对出错与系统发育推论出错等。

假设要从树中删除 3 个叶节点（已用红色着色，见图 2.4A）。drop.tip() 函数不仅可以用于删除指定的叶节点，还可以用于自动更新树图（见图 2.4B）。在新的树图中，所有的关联数据都会被保留。

```
f <- system.file("extdata/NHX", "phyldog.nhx", package="treeio")
nhx <- read.nhx(f)
to_drop <- c("Physonect_sp_@2066767",
             "Lychnagalma_utricularia@2253871",
             "Kephyes_ovata@2606431")
p1 <- ggtree(nhx) + geom_tiplab(aes(color = label %in% to_drop)) +
  scale_color_manual(values=c("black", "red")) + xlim(0, 0.8)
```

```
nhx_reduced <- drop.tip(nhx, to_drop)
p2 <- ggtree(nhx_reduced) + geom_tiplab() + xlim(0, 0.8)
plot_list(p1, p2, ncol=2, tag_levels = "A")
```

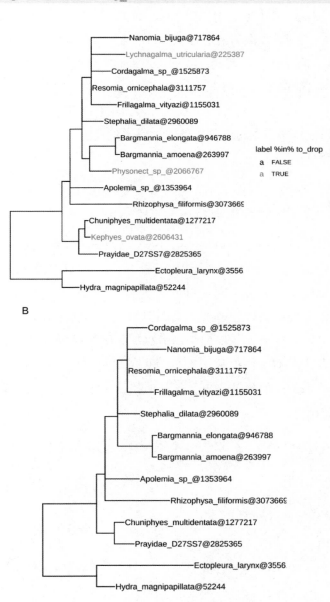

图 2.4 从树中删除叶节点

含有 3 个需要移除的叶节点（红色）的原始树（A）；删除选定叶节点之后更新的树（B）。

2.5.2 通过叶节点标签对树取子集

有时，一棵树可能会大到让我们很难专注于感兴趣的部分。treeio 包[1]中的 tree_subset() 函数用于提取树的部分子集，并且仍然保留相应的拓扑结构。图 2.5A 中的 beast_tree 略显拥挤。此时，我们可以通过把图变高从而拥有更多空间容纳叶节点标签（类似于使用 FigTree 中的 Expansion 滑块），也可以把标签文本缩小。如果树变得更大（有成百上千个节点），这个方法就没那么好用了。特别是当我们只对树在特定叶节点附近的部分感兴趣时，肯定也不会想在这样一棵大树中寻找感兴趣的物种。

假设我们对树中的叶节点 A/Swine/HK/168/2012 感兴趣（见图 2.5A），并想查看这个节点的直系亲属。

此时我们可以使用 tree_subset() 函数只查看树中感兴趣的部分。在默认情况下，当使用 tree_subset() 函数在内部调用 groupOTU() 函数时，可将指定的叶节点与其他叶节点分为不同的组（见图 2.5B）。此外，在对树取子集之后，枝长与相关数据并不会发生改变（见图 2.5C）。在默认情况下，子树中根节点的位置将被设定为零点，并根据与该节点的相对距离设置其他节点的位置。如果想要将其改为相对于原始树中的根节点的距离，则可以将 root.position 参数设置为子树的根分支的长度，也就是从原始树根节点到子树根节点枝长的总和（见图 2.5D、图 2.5E）。

```
beast_file <- system.file("examples/MCC_FluA_H3.tree",
package="ggtree")
beast_tree <- read.beast(beast_file)

p1 = ggtree(beast_tree) +
  geom_tiplab() +  xlim(0, 40) + theme_tree2()

tree2 = tree_subset(beast_tree, "A/Swine/HK/168/2012", levels_back=4)
 p2 <- ggtree(tree2, aes(color=group)) +
  scale_color_manual(values = c("black", "red"), guide = 'none') +
  geom_tiplab() +  xlim(0, 4) + theme_tree2()

p3 <- p2 +
```

```
  geom_point(aes(fill = rate), shape = 21, size = 4) +
  scale_fill_continuous(low = 'blue', high = 'red') +
  xlim(0,5) + theme(legend.position = 'right')

p4 <- ggtree(tree2, aes(color=group),
          root.position = as.phylo(tree2)$root.edge) +
  geom_tiplab() + xlim(18, 24) +
  scale_color_manual(values = c("black", "red"), guide = 'none') +
  theme_tree2()

p5 <- p4 +
  geom_rootedge() + xlim(0, 40)

plot_list(p1, p2, p3, p4, p5,
          design="AABBCC\nAADDEE", tag_levels='A')
```

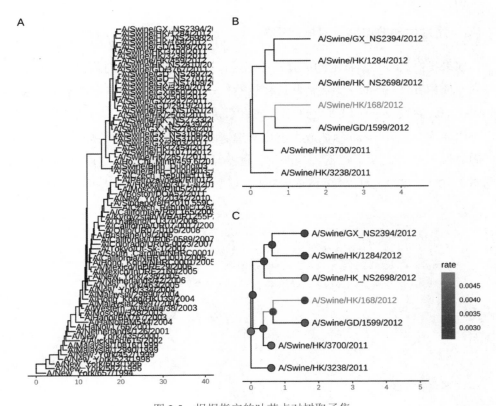

图 2.5　根据指定的叶节点对树取子集

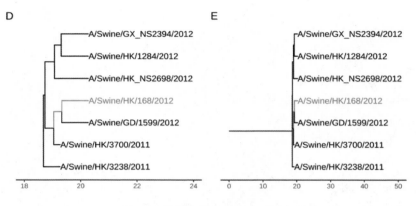

图 2.5　根据指定的叶节点对树取子集（续）

原始树（A）；子树（B）；带有数据的子树（C）；可视化相对于原始位置的子集树，没有根枝长（D）；有根枝长（E）。

2.5.3　通过内部节点编号对树取子集

如果我们只对某个特定的进化枝感兴趣，则可以将对应的内部节点编号作为输入，使用 tree_subset() 函数将此进化枝看作一个整体，追溯一定水平（可通过 levels_back 参数设置）以查看进化枝的直系亲属（见图 2.6A 和图 2.6B）。我们也可以使用 tree_subset() 函数放大进化树选定的部分，并将整棵树与选定的部分一起绘制出来，以此来探索一棵非常大的树（类似于 ape::zoom()）（见图 2.6C 和图 2.6D）。我们也可以使用 viewClade() 函数仅对特定进化枝进行可视化。

```
clade <- tree_subset(beast_tree, node=121, levels_back=0)
clade2 <- tree_subset(beast_tree, node=121, levels_back=2)
p1 <- ggtree(clade) + geom_tiplab() + xlim(0, 5)
p2 <- ggtree(clade2, aes(color=group)) + geom_tiplab() +
    xlim(0, 8) + scale_color_manual(values=c("black", "red"))

library(ape)
library(tidytree)
library(treeio)

data(chiroptera)

nodes <- grep("Plecotus", chiroptera$tip.label)
chiroptera <- groupOTU(chiroptera, nodes)
```

```
clade <- MRCA(chiroptera, nodes)
x <- tree_subset(chiroptera, clade, levels_back = 0)

p3 <- ggtree(chiroptera, aes(colour = group)) +
  scale_color_manual(values=c("black", "red")) +
  theme(legend.position = "none")
p4 <- ggtree(x) + geom_tiplab() + xlim(0, 5)
plot_list(p1, p2, p3, p4,
  ncol=2, tag_levels = 'A')
```

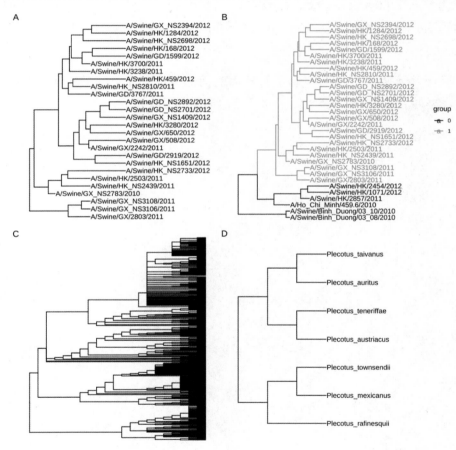

图 2.6 指定进化枝的子集树

提取选定的进化枝（A）；提取进化枝并向前追溯，以查看其直系亲属（B）；对整棵树（C）及其中的特定进化枝进行可视化（D）。

2.6 操作树数据以进行可视化

我们可以通过 ggtree[2] 对树进行可视化。虽然 ggtree 实现了很多对含有数据的树进行可视化探索的方法，但是我们可能想要呈现一些 ggtree 不支持直接实现的效果。这时，我们可以直接对节点的坐标进行操作，以实现对可视化结果的调整。使用 ggtree 能够轻松地实现这种操作，我们可以使用 fortify() 函数，该函数会在内部调用 tidytree::as_tibble() 函数来将树转换为一个简洁数据框，并添加多个含有可视化的坐标信息的列（如 x、y、branch、angle 等）。我们也可以通过 ggtree(tree)$data 来访问这些数据。

在下面示例中，我们绘制了两棵面对面的系统发育树，其结果类似于由 ape::cophyloplot() 函数生成的图，如图 2.7 所示。

```
library(dplyr)
library(ggtree)

set.seed(1024)
x <- rtree(30)
y <- rtree(30)
p1 <- ggtree(x, layout='roundrect') + 
  geom_hilight(
       mapping=aes(subset = node %in% c(38, 48, 58, 36),
                   node = node,
                   fill = as.factor(node)
                   )
  ) +
  labs(fill = "clades for tree in left" )

p2 <- ggtree(y)

d1 <- p1$data
d2 <- p2$data

## 翻转 x 轴，并通过设定 offset 来保证第二棵树会出现在第一棵树的右侧
d2$x <- max(d2$x) - d2$x + max(d1$x) + 1

pp <- p1 + geom_tree(data=d2, layout='ellipse') + 
  ggnewscale::new_scale_fill() +
```

```r
  geom_hilight(
      data = d2,
      mapping = aes(
          subset = node %in% c(38, 48, 58),
          node=node,
          fill=as.factor(node))
  ) +
  labs(fill = "clades for tree in right" )

dd <- bind_rows(d1, d2) %>%
  filter(!is.na(label))

pp + geom_line(aes(x, y, group=label), data=dd, color='grey') +
    geom_tiplab(geom = 'shadowtext', bg.colour = alpha('firebrick', .5)) +
    geom_tiplab(data = d2, hjust=1, geom = 'shadowtext',
                bg.colour = alpha('firebrick', .5))
```

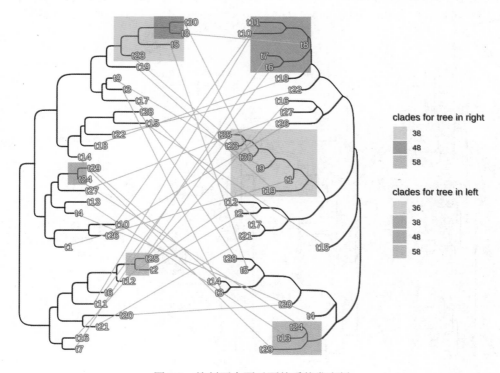

图 2.7 绘制两个面对面的系统发育树

使用 ggtree() 函数先在左侧绘制一棵树，再通过 geom_tree() 图层函数在右侧绘制另一棵树。我们可以调整两棵树的相对位置，或者独立地向每棵树添加图层（如叶节点标签或对进化枝进行突出显示）。

我们可以轻松地在同一张图中绘制多棵进化树，并连接其中的分类单元。例如，我们可以将由流感病毒所有内部基因片段构建的进化树绘制在一起，并将不同树之间属于同一个病毒株的基因连接起来[8]。图 2.8 演示了如何通过组合多个 geom_tree() 图层函数来生成多棵进化树。

```
z <- rtree(30)
d2 <- fortify(y)
d3 <- fortify(z)
d2$x <- d2$x + max(d1$x) + 1
d3$x <- d3$x + max(d2$x) + 1

dd = bind_rows(d1, d2, d3) %>%
  filter(!is.na(label))

p1 + geom_tree(data = d2) + geom_tree(data = d3) + geom_tiplab(data=d3) +
  geom_line(aes(x, y, group=label, color=node < 15), data=dd, alpha=.3)
```

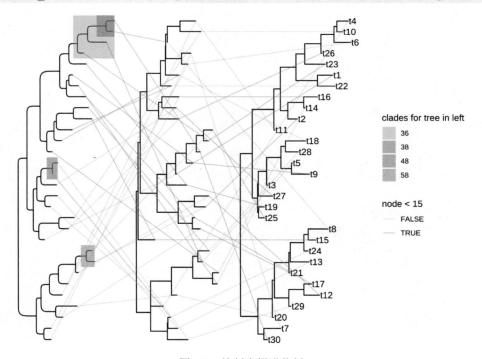

图 2.8　绘制多棵进化树

我们使用 ggtree() 函数先绘制出一棵树，再通过 geom_tree() 图层函数添加多个进化树图层。

2.7 总结

treeio 包可用于将各式各样的系统发育相关数据导入 R 中。由于系统发育树是以一种便于进行计算处理的方式存储的，因此并不是很便于我们以人类的思维理解，也使得必须要有相当的专业知识的人才能够对进化树数据进行操作及探索。为此，tidytree 包提供了一个用于探索树数据的整洁接口。ggtree 也提供了一组实用工具，通过使用图形语法对树数据进行可视化及探索。这一整套软件包不仅让普通用户也能轻松地与树数据进行交互，还能让我们整合来自不同来源（例如实验结果或分析结果）的系统发育相关数据，从而使得我们能进行相应的整合研究及比较研究。此外，这套软件包还将系统发育分析带入了 tidyverse 中，而这也将我们对于系统发育数据的处理带入了一个全新的境界。

2.8 本章练习题

1. 如何实现 phylo 与 tbl_tree 对象、treedata 对象之间的互相转换。

2. 随机生成一棵树，并访问其给定一叶节点的祖先节点及内部节点的后代节点。

3. 随机生成一棵树并获取任意一个内部节点的子集树；获取任意一个叶节点并向前追溯一个水平（父节点），以此取子集树并通过 ggtree 对其进行可视化。

参考文献

[1] Wang L, Lam T T, Xu S, et al. treeio: an R package for phylogenetic tree input and output with richly annotated and associated data[J]. Molecular Biology and Evolution, 2020,37(2):599-603.

[2] Yu G, Smith D K, Zhu H, et al. ggtree: an R package for visualization and annotation of phylogenetic trees with their covariates and other associated data[J]. Methods in Ecology and Evolution, 2016, 8(1): 28-36.

[3] Xu S, Dai Z, Guo P, et al. ggtreeExtra: Compact visualization of richly annotated phylogenetic data[J]. Mol Biol Evol, 2021, 38(9): 4039-4042.

[4] Paradis E, Claude J, Strimmer K. APE: analyses of phylogenetics and evolution in R language[J]. Bioinformatics, 2004, 20(2): 289-290.

[5] Yu G, Lam T T, Zhu H, et al. Two methods for mapping and visualizing associated data on phylogeny using ggtree[J]. Mol Biol Evol, 2018, 35(12): 3041-3043.

[6] Liang H, Lam T T, Fan X, et al. Expansion of genotypic diversity and establishment of 2009 H1N1 pandemic-origin internal genes in pigs in China[J]. J Virol, 2014, 88(18): 10864-10874.

[7] Czech L, Huerta-Cepas J, Stamatakis A. A critical review on the use of support values in tree viewers and bioinformatics toolkits[J]. Mol Biol Evol, 2017, 34(6): 1535-1542.

[8] Venkatesh D, Poen M, Bestebroer T, et al. Avian influenza viruses in wild birds: virus evolution in a multihost ecosystem[J]. Journal of Virology, 2018, 92(15).

第3章　导出含有数据的树

3.1　简介

treeio 包[1]支持解析多种系统发育树的文件格式，如含有进化证据的软件输出文件。有些格式只是简单的日志文件（如 PAML 和 r8s 的输出文件），而有些则是非标准格式（如引入了方括号的 BEAST 与 MrBayes 的输出文件，其中的方括号在标准 Nexus 文件中被保留下来，用于存储进化推论）。我们可以使用 treeio 解析这些文件，提取其中的系统发育树并将关联数据映射到树的结构上。仅导出树结构是比较简单的，用户可以先使用 treeio 中的 as.phyo() 函数将 treedata 对象转换为 phylo 对象，再使用 ape 包[2]中的 write.tree() 函数或 write.nexus() 函数将树结构导出为 Newick 文件或 Nexus 文件。这种方法在将非标准格式转换为标准格式，以及从软件输出文件（比如日志文件）中提取树时非常有用。

相比之下，导出含有相关数据的树仍然十分具有挑战性。这些相关联的数据可以是从分析程序的结果中解析出来的，也可以是从其他外部来源（如表型数据、实验数据及临床数据）获得的。目前最大的阻碍还是缺少一个用于储存含数据的树的标准格式。NeXML[3]可能是目前最灵活的格式，但它仍未得到各个软件的广泛支持。该领域的大多数分析程序还是广泛依赖于 Newick 字符串及 Nexus 格式。虽然 BEAST Nexus 格式可能并不是最好的解决方案，但在当前是存储异构关联数据的好方法。这种格式的优势在于所有的注释元素都存储于方括号中，而方括号本来是用于存储注解信息的。这样一来，现有的支持读取标准 Nexus 文件的程序也可以通过将其中的注释元素作为注解信息忽略，来读取 BEAST Nexus 文件。

3.2 将树数据导出为 BEAST Nexus 格式的文件

3.2.1 软件输出文件的导出与转换

treeio 包[1]提供了 write.beast() 函数，用于将 treedata 对象导出为 BEAST Nexus 格式的文件[4]。我们使用 treeio 可以很轻松地将 treeio 能够读取的软件输出文件转换为 BEAST Nexus 文件（参见第 1 章）。

下面是将 NHX 文件转换为 BEAST Nexus 文件的示例。

```
nhxfile <- system.file("extdata/NHX", "phyldog.nhx", package="treeio")
nhx <- read.nhx(nhxfile)
# write.beast(nhx, file = "phyldog.tree")
write.beast(nhx)
```

```
#NEXUS
[R-package treeio, Thu Oct 14 11:24:19 2021]

BEGIN TAXA;
    DIMENSIONS NTAX = 16;
    TAXLABELS
        Prayidae_D27SS7@2825365
        Kephyes_ovata@2606431
        Chuniphyes_multidentata@1277217
        Apolemia_sp_@1353964
        Bargmannia_amoena@263997
        Bargmannia_elongata@946788
        Physonect_sp_@2066767
        Stephalia_dilata@2960089
        Frillagalma_vityazi@1155031
        Resomia_ornicephala@3111757
        Lychnagalma_utricularia@2253871
        Nanomia_bijuga@717864
        Cordagalma_sp_@1525873
        Rhizophysa_filiformis@3073669
        Hydra_magnipapillata@52244
        Ectopleura_larynx@3556167
    ;
END;
```

```
BEGIN TREES;
    TRANSLATE
        1    Prayidae_D27SS7@2825365,
        2    Kephyes_ovata@2606431,
        3    Chuniphyes_multidentata@1277217,
        4    Apolemia_sp_@1353964,
        5    Bargmannia_amoena@263997,
        6    Bargmannia_elongata@946788,
        7    Physonect_sp_@2066767,
        8    Stephalia_dilata@2960089,
        9    Frillagalma_vityazi@1155031,
        10   Resomia_ornicephala@3111757,
        11   Lychnagalma_utricularia@2253871,
        12   Nanomia_bijuga@717864,
        13   Cordagalma_sp_@1525873,
        14   Rhizophysa_filiformis@3073669,
        15   Hydra_magnipapillata@52244,
        16   Ectopleura_larynx@3556167
    ;
    TREE * UNTITLED = [&R] (((1[&Ev=S,ND=0,S=58]:0.0682841,(2[&Ev=S,
ND=1,
S=69]:0.0193941,3[&Ev=S,ND=2,S=70]:0.0121378)[&Ev=S,ND=3,S=60]:0.0217782)
[&Ev=S,ND=4,S=36]:0.0607598,((4[&Ev=S,ND=9,S=31]:0.11832,(((5[&Ev=S,ND=10,
S=37]:0.0144549,6[&Ev=S,ND=11,S=38]:0.0149723)[&Ev=S,ND=12,S=33]: 0.0925388,
7[&Ev=S,ND=13,S=61]:0.077429)[&Ev=S,ND=14,S=24]:0.0274637,(8[&Ev=S,
ND=15,
S=52]:0.0761163,((9[&Ev=S,ND=16,S=53]:0.0906068,10[&Ev=S,ND=17,S=54]
:1e-06)
[&Ev=S,ND=18,S=45]:1e-06,((11[&Ev=S,ND=19,S=65]:0.120851,12[&Ev=S,ND=20,
S=71]:0.133939)[&Ev=S,ND=21,S=56]:1e-06,13[&Ev=S,ND=22,S=64]:0.0693814)
[&Ev=S,ND=23,S=46]:1e-06)[&Ev=S,ND=24,S=40]:0.0333823)[&Ev=S,ND=25,
S=35]:
1e-06)[&Ev=D,ND=26,S=24]:0.0431861)[&Ev=S,ND=27,S=19]:1e-06,14[&Ev=S,
ND=28,
S=26]:0.22283)[&Ev=S,ND=29,S=17]:0.0292362)[&Ev=D,ND=8,S=17]:0.185603,
(15[&Ev=S,ND=5,S=16]:0.0621782,16[&Ev=S,ND=6,S=15]:0.332505)[&Ev=S,ND=7,
S=12]:0.185603)[&Ev=S,ND=30,S=9];
END;
```

下面是将 CODEML 输出文件转换为 BEAST Nexus 输出文件的另一个示例。

```
mlcfile <- system.file("extdata/PAML_Codeml", "mlc", package="treeio")
ml <- read.codeml_mlc(mlcfile)
# write.beast(ml, file = "codeml.tree")
```

```
write.beast(ml)  # 此处并未展示输出结果
```

有的软件并不支持解析这些输出文件，我们可以通过数据转换将输出文件格式转换为软件支持的格式。例如，我们可以将 NHX 文件转换为 FigTree 能识别的 BEAST Nexus 文件，再使用 FigTree 利用其中的 NHX 标签为树着色（见图 3.1A）；或者转换 CODEML 的输出文件格式，再使用 FigTree 中的 dN/dS、dN 或 dS 为树着色（见图 3.1B）。在进行数据转换之前，Figtree 是无法打开这些文件的。treeio 的数据转换功能使数据可供其他软件工具使用，扩展了这些软件工具的应用范围。

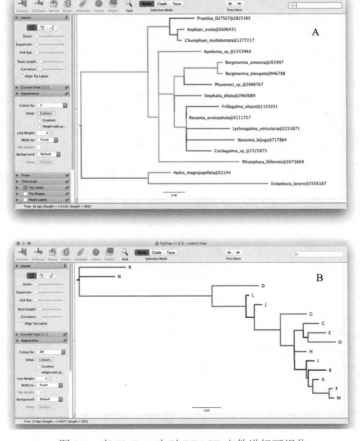

图 3.1　在 FigTree 中对 BEAST 文件进行可视化

FigTree 并不支持直接可视化 NHX 文件（A）和 CODEML 输出文件（B）。treeio 可以先将这些文件转换为 FigTree 兼容的 BEAST Nexus 文件, 再通过 FigTree 打开文件并对树与注释数据一同进行可视化。

3.2.2 将树与外部数据结合

我们使用 tidytree 和 treeio 提供的工具可以轻松地将外部数据链接到相应的系统发育树上。同时, write.beast() 函数能够帮助我们将这些含有外部数据的树导入单个树文件中。

```
phylo <- as.phylo(nhx)
## 为输出树文本留出空间
phylo$edge.length <- round(phylo$edge.length, 2)

## 输出 Newick 文本
write.tree(phylo)
```

```
(((Prayidae_D27SS7@2825365:0.07,(Kephyes_ovata@2606431:0.02,
Chuniphyes_multidentata@1277217:0.01):0.02):0.06,((Apolemia_sp_
@1353964:0.12,
(((Bargmannia_amoena@263997:0.01,Bargmannia_elongata@946788:0.01):0.09,
Physonect_sp_@2066767:0.08):0.03,(Stephalia_dilata@2960089:0.08,
((Frillagalma_vityazi@1155031:0.09,Resomia_ornicephala@3111757:0):0,
((Lychnagalma_utricularia@2253871:0.12,Nanomia_bijuga@717864:0.13):0,
Cordagalma_sp_@1525873:0.07):0):0.03):0.04):0,Rhizophysa_
filiformis@3073669:
0.22):0.03):0.19,(Hydra_magnipapillata@52244:0.06,
Ectopleura_larynx@3556167:0.33):0.19);
```

```
N <- Nnode2(phylo)
fake_data <- tibble(node = 1:N, fake_trait = round(rnorm(N), 2),
                    another_trait = round(runif(N), 2))
fake_tree <- full_join(phylo, fake_data, by = "node")
# write.beast(fake_tree)

## 为了节省空间, 此处仅展示子树
fake_tree2 = tree_subset(fake_tree, node=27, levels_back=0)
write.beast(fake_tree2)
```

```
#NEXUS
[R-package treeio, Tue Nov 16 10:13:32 2021]
```

```
BEGIN TAXA;
    DIMENSIONS NTAX = 5;
    TAXLABELS
        Frillagalma_vityazi@1155031
        Resomia_ornicephala@3111757
        Lychnagalma_utricularia@2253871
        Nanomia_bijuga@717864
        Cordagalma_sp_@1525873
    ;
END;
BEGIN TREES;
    TRANSLATE
        1   Frillagalma_vityazi@1155031,
        2   Resomia_ornicephala@3111757,
        3   Lychnagalma_utricularia@2253871,
        4   Nanomia_bijuga@717864,
        5   Cordagalma_sp_@1525873
    ;
    TREE * UNTITLED = [&R] ((1[&fake_trait=-1.42,another_trait=0.17]: 0.09,
2[&fake_trait=0.19,another_trait=0.04]:0)[&fake_trait=0.85,
another_trait=0.56]:0,(5[&fake_trait=0.22,another_trait=0.73]:0.07,
(3[&fake_trait=0.02,another_trait=0.29]:0.12,4[&fake_trait=-1.29,
another_trait=0.35]:0.13)[&fake_trait=-0.33,another_trait=0.88]:0)
[&fake_trait=0.27,another_trait=0.94]:0):0.29;
END;
```

在与树合并后，存储在 fake_data 中的 fake_trait 和 another_trait 将被关联到 phylo 树，并存储在 treedata 对象 fake_tree 中。write.beast() 函数用于将包含关联数据的树导入单个 BEAST Nexus 文件中，其中的关联数据可以使用 ggtree 或 FigTree 对树进行可视化。

3.2.3 合并不同来源的树数据

我们不仅可以将 Newick 树文本与关联数据相结合，也可以将从软件输出文件中得到的树数据与外部数据相结合，或者合并不同的树对象（参见第 2 章）。

下面的代码展示了如何将树与数据结合。

```
## 把树对象和数据结合起来
tree_with_data <- full_join(nhx, fake_data, by = "node")
tree_with_data
```

```
## 'treedata' S4 object that stored information
## of
##  '/home/ygc/R/library/treeio/extdata/NHX/phyldog.nhx'.
##
## ...@ phylo:
##
## Phylogenetic tree with 16 tips and 15 internal nodes.
##
## Tip labels:
##    Prayidae_D27SS7@2825365, Kephyes_ovata@2606431,
## Chuniphyes_multidentata@1277217,
## Apolemia_sp_@1353964, Bargmannia_amoena@263997,
## Bargmannia_elongata@946788, ...
##
## Rooted; includes branch lengths.
##
## with the following features available:
##    'Ev', 'ND', 'S', 'fake_trait', 'another_trait'.
```

下面的代码展示了如何合并两个树对象。

```
## 合并两个树对象
tree2 <- merge_tree(nhx, fake_tree)
identical(tree_with_data, tree2)
```

```
## [1] TRUE
```

在与不同来源的数据合并后，含有关联数据的树可以被导入单个文件中。

```
outfile <- tempfile(fileext = ".tree")
write.beast(tree2, file = outfile)
```

输出的 BEAST Nexus 文件可以通过 read.beast() 函数导入 R 中，并且所有关联数据都可以通过 ggtree[5] 对树进行注释。

```
read.beast(outfile)
```

```
## 'treedata' S4 object that stored information
## of
##  '/tmp/Rtmp13jbns/file26455ef94978.tree'.
##
## ...@ phylo:
```

```
## 
## Phylogenetic tree with 16 tips and 15 internal nodes.
## 
## Tip labels:
##     Prayidae_D27SS7@2825365, Kephyes_ovata@2606431,
## Chuniphyes_multidentata@1277217,
## Apolemia_sp_@1353964, Bargmannia_amoena@263997,
## Bargmannia_elongata@946788, ...
## 
## Rooted; includes branch lengths.
## 
## with the following features available:
##     'another_trait', 'Ev', 'fake_trait', 'ND', 'S'.
```

3.3　将树数据导出为 jtree 格式的文件

　　treeio 包[1]中的 write.beast() 函数用于将 treedata 对象导入 BEAST Nexus 文件中，这对于转换文件格式、将树与数据相结合及合并不同来源的树数据都非常有用。treeio 包中的 read.beast() 函数用于解析 write.beast() 函数的输出文件。有了 treeio 包，R 社区就能够处理 BEAST Nexus 文件与处理树数据。但是在其他编程语言中仍缺少用于解析 BEAST Nexus 文件的库或包。

　　JSON（JavaScript Object Notation）是一种轻量级的数据交换格式，并在几乎所有的现代编程语言中都得到了广泛支持。为了便于在其他编程语言中导入含有数据的树，treeio 包支持以 jtree 格式导出包含数据的树。jtree 格式基于 JSON 构建，并可以使用任何支持 JSON 的语言轻松地进行解析。

```
# write.jtree(tree2)

# 为了节省空间，此处仅展示子树
tree3 <- tree_subset(tree2, node=24, levels_back=0)
write.jtree(tree3)
```

```
{
    "tree": "(Physonect_sp_@2066767:0.077429{3},(Bargmannia_
amoena@263997:0.0144549
{1},Bargmannia_
elongata@946788:0.0149723{2}):0.0925388{5}):0.28549{4};",
    "data":[
```

```
  {
    "edge_num": 1,
    "Ev": "S",
    "ND": 10,
    "S": 37,
    "fake_trait": -0.69,
    "another_trait": 0.42
  },
  {
    "edge_num": 2,
    "Ev": "S",
    "ND": 11,
    "S": 38,
    "fake_trait": -0.95,
    "another_trait": 0.38
  },
  {
    "edge_num": 3,
    "Ev": "S",
    "ND": 13,
    "S": 61,
    "fake_trait": 0.59,
    "another_trait": 0.65
  },
  {
    "edge_num": 4,
    "Ev": "S",
    "ND": 14,
    "S": 24,
    "fake_trait": -0.69,
    "another_trait": 0.06
  },
  {
    "edge_num": 5,
    "Ev": "S",
    "ND": 12,
    "S": 33,
    "fake_trait": -0.58,
    "another_trait": 0.4
  }
],
    "metadata": {"info": "R-package treeio", "data": "Tue Nov 16 10:21:20 2021"}
}
```

jtree 文件格式基于 JSON 构建，并且可以被 JSON 解析器解析。

```
jtree_file <- tempfile(fileext = '.jtree')
write.jtree(tree2, file = jtree_file)
jsonlite::fromJSON(jtree_file)
```

```
$tree
[1] "(Physonect_sp_@2066767:0.077429{3},(Bargmannia_
amoena@263997:0.0144549{1},
Bargmannia_elongata@946788:0.0149723{2}):0.0925388{5}):0.28549{4};"

$data
  edge_num Ev ND  S fake_trait another_trait
1        1  S 10 37      -0.69          0.42
2        2  S 11 38      -0.95          0.38
3        3  S 13 61       0.59          0.65
4        4  S 14 24      -0.69          0.06
5        5  S 12 33      -0.58          0.40


$metadata
$metadata$info
[1] "R-package treeio"

$metadata$data
[1] "Tue Nov 16 10:24:34 2021"
```

我们通过 treeio 包中的 read.jtree() 函数可以将 jtree 文件作为 treedata 对象直接导入 R 中。

```
read.jtree(jtree_file)
```

```
## 'treedata' S4 object that stored information
## of
##   '/tmp/Rtmp13jbns/file2645630f9a5.jtree'.
##
## ...@ phylo:
##
## Phylogenetic tree with 16 tips and 15 internal nodes.
##
## Tip labels:
##   Prayidae_D27SS7@2825365, Kephyes_ovata@2606431,
## Chuniphyes_multidentata@1277217,
```

```
## Apolemia_sp_@1353964, Bargmannia_amoena@263997,
## Bargmannia_elongata@946788, ...
##
## Rooted; includes branch lengths.
##
## with the following features available:
##   'Ev', 'ND', 'S', 'fake_trait', 'another_trait'.
```

3.4 总结

系统进化树的关联数据通常被存储在不同的文件中，需要用户具有相应的专业知识才能将数据映射到树结构上。由于缺少存储和表示系统发育树与其关联数据的统一标准，科研工作者很难访问系统发育数据并将其整合到研究中。treeio 包提供了多种函数来导入系统发育树及其关联数据，这些数据可以通过多个来源（如常见分析软件的输出结果，或者实验数据、临床数据、元数据等外部数据）获得。树及其关联数据可以以 BEAST Nexus 或 jtree 格式导入单个文件中，这些输出文件又可以通过 treeio 解析到 R 中，并轻松获取其中的信息。treeio 包提供的输入及输出工具为系统发育数据的整合，以及下游的比较分析和可视化奠定了基础，让我们能够对树与不同来源的关联数据进行整合，拓展了系统发育分析在其他学科中的应用。

3.5 本章练习题

1. 如何将 BEAST 文件和 NHX 合并，并将其导出为 BEAST 文件与 Newick 文件？

2. 如何从 R 中导入、导出 jtree 对象？

参考文献

[1] Wang L, Lam T T, Xu S, et al. treeio: an R package for phylogenetic tree input and output with richly annotated and associated data[J]. Molecular Biology and Evolution, 2020,37(2):599-603.

[2] Paradis E, Claude J, Strimmer K. APE: analyses of phylogenetics and evolution in R

language[J]. Bioinformatics, 2004, 20(2): 289-290.

[3] Vos R A, Balhoff J P, Caravas J A, et al. NeXML: rich, extensible, and verifiable representation of comparative data and metadata[J]. Syst Biol, 2012, 61(4): 675-689.

[4] Bouckaert R, Heled J, Kuhnert D, et al. BEAST 2: a software platform for Bayesian evolutionary analysis[J]. PLoS Comput Biol, 2014, 10(4): e1003537.

[5] Yu G, Smith D K, Zhu H, et al. ggtree: an R package for visualization and annotation of phylogenetic trees with their covariates and other associated data[J]. Methods in Ecology and Evolution, 2016, 8(1): 28-36.

第 2 篇
树数据的可视化及注释

第 4 章　系统发育树可视化

4.1　简介

目前有很多为进化树可视化设计的软件包或网页版的工具，如 TreeView[1]、FigTree、TreeDyn[2]、Dendroscope[3]、EvolView[4] 及 iTOL[5] 等。其中，只有 FigTree、TreeDyn 和 iTOL 等少数软件支持对树的注释操作，如为分类单元着色或根据树的特征高亮相应的进化枝等，这些预先设定好的注释功能也只适用于特定类型的系统进化数据。随着进化树在多学科联合分析中的应用越来越广，将不同种类的系统进化协变量和其他不同来源的进化树相关数据与进化树联合起来进行可视化和后续分析的需求也越来越大。例如，流感病毒具有广泛的宿主类型，多样化且动态变化的基因类型及特征性的传播行为，对于流感病毒来说是本质性的，并且大多数与病毒的进化有关。因此，除了专注于特定分析方式或数据类型的独立应用程序，研究分子进化的研究人员还需要一个稳健且可编程的平台。该平台允许对系统发育树上不同特征的数据（原始数据或来源于分析结果）进行高度整合与可视化，以识别它们的关联与模式。

为了填补这一空白，我们开发了 ggtree 包[6]，这是一个发布在 Bioconductor 项目[7] 下的 R[8] 软件包。ggtree 构建于 treedata 对象之上，并基于图形语法[9] 开发的 ggplot2 包[10] 可用于绘制进化树。

R 在系统发育学研究中的应用越来越广泛，但是目前仍缺少一款功能全面，且专门为进化树的可视化及注释，特别是复杂数据整合设计的软件包。大多数的系统发育学相关 R 软件包都只是专注于特定的统计学分析，而没有把目光放到进化树的可视化，以及使用更广泛的系统发育相关数据对树进行注释上。当然，还是有诸如 ape[11] 和 phytools[12] 等包，基于 R 的 Base 绘图系统实现了进化树的可视化及注释，特别是 ape 包，可以说是系统发育分析及数据处理最基本的包之一。由于 Base 绘图系统相对来说较难进行拓展，限制了绘制出的进化树图的多样性。而 OutbreakTools[13] 与 phyloseq[14] 则对 ggplot2 包进行了拓展，实现了系统发育树的可视化。ggplot2 绘图系统相比于 base，支持更为便捷的自定义功能。但这两个包分别是针对流行病学与微生物组数据开发的，并不能为进化树的可视化与注释提供一个更为通用的解决方案。相比之下，ggtree 包在继承了 ggplot2 包诸多特性的同时，还支持通过导入来自不同来源的进化树关联数据（参见第 1 章及 treeio 文章[15]），生成并自由组合多个注释图层（参见第 5 章）来构建非常复杂的树图。

4.2 使用 ggtree 包对系统发育树进行可视化

ggtree 包[6] 是专门为了使用不同种类及不同来源的相关数据注释进化树设计的。这些数据可以由用户提供或由分析程序生成，其中可能包含进化速率、祖先序列等与真实样本中的分类单元内部节点（表示假定祖先物种/菌株）或分支（表示进化时间过程）的相关信息[15]。例如，这些数据可能是病毒基因树中的禽流感病毒样本地理位置信息（由调研位置得知）、祖先节点信息（由系统地理学推断得知）。

ggtree 包支持 ggplot2 包的图形语言，继承了其高度自定义的特性，又具有直观且灵活的特点。值得一提的是，ggplot2 包本身并不支持对树形结构提供底层的几何对象或其他的支持，因此 ggtree 包在这方面是一个非常实用的拓展。除了 ggtree 包，还有 OutbreakTools 和 phyloseq 两个系统发育相关的 R 包也是基于 ggplot2 包开发的，但它们并不支持 ggplot2 包语法中最有价值的部分——添加注释层，这会导致一些操作上的不便。例如，如果想要绘制一棵没有添加分类单元标签的树，那么对于一般的 R 用户来说，由于对 OutbreakTools 包和 phyloseq 包的基础架构并不了解，想要使用它们为树添加分类单元标签是非常困难的。而 ggtree 包则是通过 geom_tree() 图层函数来对树结构进行可视化的，从而拓展了 ggplot2 包的功能。因此相比之下，使用 ggtree 包能更加轻松地绘制进化树，只需通过 ggplot(tree_object) + geom_tree() + theme_tree() 或 ggtree(tree_object) 便可以实现，后续则可以通过"+"操作符一层一层添加注释图层。为了使树的可视化变得更加便利，ggtree 包提供了一些用于进化树注释的图层函数，包括添加分支标尺图例的图层函数 geom_treescale()（可用于显示遗传距离、分歧时间等）、显示枝长不确定性的图层函数 geom_range()（可用于显示置信区间或极差等）、添加分类单元标签的图层函数 geom_tiplab()、分别在内/外部节点位置添加符号的图层函数 geom_nodepoint() 与 geom_tippoint()、使用矩形对进化枝进行高亮显示的图层函数 geom_hilight()、使用线条及文本标签注释选定进化枝的图层函数 geom_cladelab() 等。

要想对系统发育树进行可视化，我们首先需要利用 treeio 包将树文件解析至 R 中。treeio 包可以将不同软件输出的、包含各类注释数据的结果解析至 S4 类系统发育数据对象中（参见第 1 章）。ggtree 包主要是利用这些 S4 对象来对树进行可视化及注释的。另外，其他 R 包为存储系统发育树及特定领域的关联数据定义了 S3 或 S4 类，如在 phylobase 包中定义的 phylo4 和 phylo4d、在 OutbreakTools 包中定义的 obkdata 及在 phyloseq 包中定义的 phyloseq 等。ggtree 包同样支持这些树对象的可视化，同时其中存储的注释数据也能直接被用于对进化树进行注释（参见第 9 章）。ggtree 包这种强大的兼容性能为数据与分析结果的整合带来很多便利。除此之外，ggtree 包还支持其他树形结构的可视化，包括树状图（dendrogram）和树形网络图（tree graph）。

4.2.1 基本的系统发育树的可视化

ggtree 包通过拓展 ggplot2 包 [10] 来实现对系统发育树可视化的支持，并通过 geom_tree() 图层函数来显示进化树，如图 4.1A 所示。

```
library("treeio")
library("ggtree")

nwk <- system.file("extdata", "sample.nwk", package="treeio")
tree <- read.tree(nwk)

ggplot(tree, aes(x, y)) + geom_tree() + theme_tree()
```

ggtree() 函数作为一种可视化树的快捷方式，其工作原理与上述代码完全相同。

ggtree 包汲取了 ggplot2 包的所有优点，如我们可以像使用 ggplot2 包一样更改线条的颜色、大小和类型，如图 4.1B 所示。

```
ggtree(tree, color="firebrick", size=2, linetype="dotted")
```

在默认情况下，进化树会以阶梯形的形式呈现。用户可以通过设置参数 ladderize = FALSE 来关闭这个功能（见图 4.1C，另见附录 A 中的图 A.5）。

```
ggtree(tree, ladderize=FALSE)
```

branch.length 参数用于修改分支长度的标尺。用户可以通过设置参数 branch.length = "none" 来仅查看树的拓扑结构［也就是生成分支图（cladogram）］，如图 4.1D 所示，或者使用其他数值变量来修改树的标尺（如 dN/dS）。

```
ggtree(tree, branch.length="none")
```

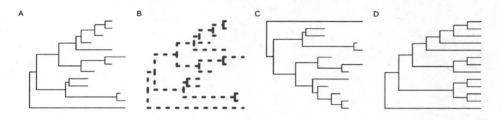

图 4.1　基本的系统发育树的可视化

默认的 ggtree 输出结果，以阶梯形呈现（A）。非变量设置（如颜色、线条粗细、线条类型等）（B）。非梯形树（C）。仅显示树拓扑而不提供枝长信息的分支图（D）。

4.2.2　系统发育树的布局

使用 ggtree 包可以非常简单地对系统发生关系进行可视化，只需要将树对象传入 ggtree() 函数即可。我们也为进化树的呈现开发了几种不同类型的布局，如图 4.2 所示，包括 rectangular（直角矩形布局，默认）、roundrect（圆角矩形布局）、ellipse（椭圆布局）、slanted（倾斜布局）等矩形布局，circular（环形布局）、fan（扇形布局，为特殊形式的环形布局）等环形布局，unrooted（无根布局）、时间尺度布局（time-scaled）、二维树布局（two-dimensional），其中 unrooted（无根布局）可通过等角（equal angle）和日光（daylight）两种算法实现。

下面是用不同布局来对进化树进行可视化的示例。

```
library(ggtree)
set.seed(2017-02-16)
tree <- rtree(50)
```

```
ggtree(tree)
ggtree(tree, layout="roundrect")
ggtree(tree, layout="slanted")
ggtree(tree, layout="ellipse")
ggtree(tree, layout="circular")
ggtree(tree, layout="fan", open.angle=120)
ggtree(tree, layout="equal_angle")
ggtree(tree, layout="daylight")
ggtree(tree, branch.length='none')
ggtree(tree, layout="ellipse", branch.length="none")
ggtree(tree, branch.length='none', layout='circular')
ggtree(tree, layout="daylight", branch.length = 'none')
```

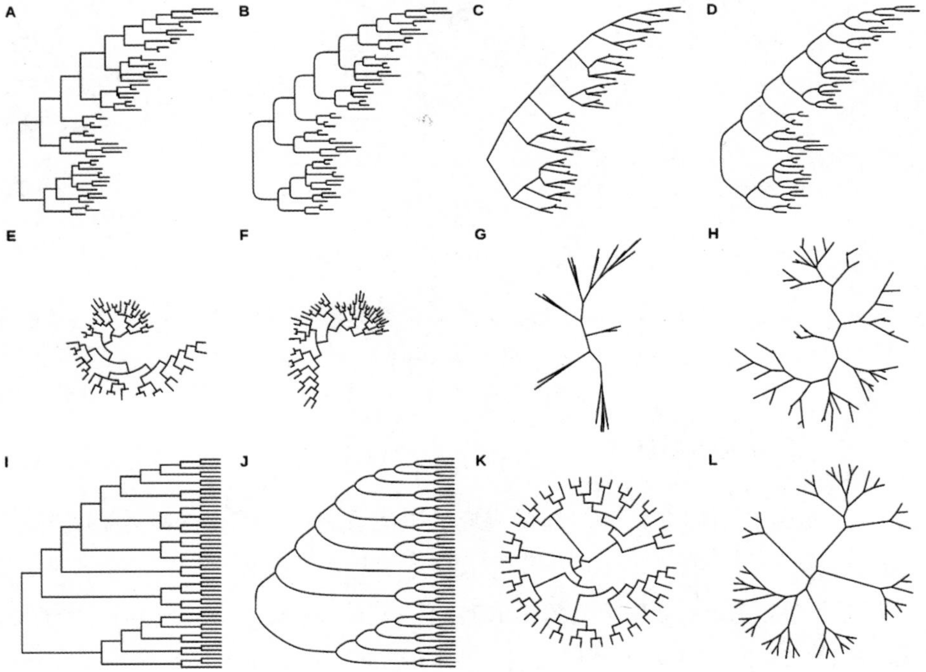

图 4.2 进化树的布局

系统发育图（phylogram）：矩形布局（A）、圆角矩形布局（B）、倾斜布局（C）、椭圆布局（D）、环形布局（E）和扇形布局（F）。

无根布局：等角算法（G）和日光算法（H）。

分支图：矩形布局（I）、椭圆布局（J）、环形布局（K）和无根布局（L）。分支图也支持倾斜布局和扇形布局，此处并未画出。

用户也可以通过修改标尺或坐标来绘制其他可能的布局,如图 4.3 所示。

```
ggtree(tree) + scale_x_reverse()
ggtree(tree) + coord_flip()
ggtree(tree) + layout_dendrogram()
ggplotify::as.ggplot(ggtree(tree), angle=-30, scale=.9)
ggtree(tree, layout='slanted') + coord_flip()
ggtree(tree, layout='slanted', branch.length='none') +
  layout_dendrogram()
ggtree(tree, layout='circular') + xlim(-10, NA)
ggtree(tree) + layout_inward_circular()
ggtree(tree) + layout_inward_circular(xlim=15)
```

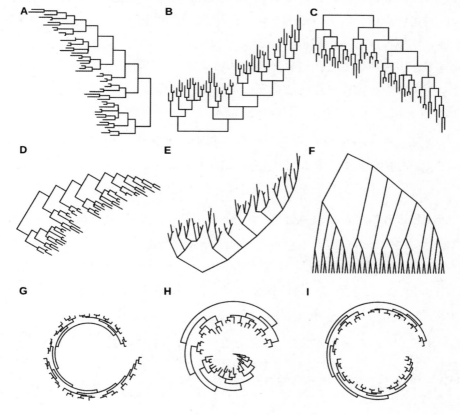

图 4.3　一些派生的进化树布局

从右到左的矩形布局(A)、自下而上的矩形布局(B)、自上而下的矩形布局(树状图布局)(C)、旋转后的矩形布局(D)、自下而上的倾斜布局(E)、自上而下的倾斜布局(分支图常见的表现形式)(F)、环形布局(G)、内向环形布局(H和I)。

系统发育图（phylogram）：ggtree 包支持以矩形布局、圆角矩形布局、倾斜布局、椭圆布局、环形布局及扇形布局来对系统发育图（默认选项，枝长根据标尺进行调整）进行可视化，如图 4.2A ～图 4.2F 所示。

无根布局（unrooted）：无根布局又被称为"放射状布局"，是通过等角算法（equal angle）或日光算法（daylight）来实现的。用户可以通过将 equal_angle（等角算法）或 daylight（日光算法）传递给 layout 参数来指定使用特定的无根布局算法对树进行可视化。等角算法由 Christopher Meacham 在 PLOTREE 中首次提出，并被收入 PHYLIP[16] 软件中。这种算法从进化树的根部开始，并根据其中的叶节点数量为每个子树分配一定角度的弧。它从根节点到叶节点进行迭代，并将分配给子树的角度再细分给下一级的子树。这种算法的速度很快，并已在许多软件包中实现。如图 4.2G 所示，等角算法有一个缺点，即叶节点倾向于聚集在一起，这会留下许多未使用的空间。日光算法从以等角算法构建的初始树开始，遍历内部节点并调节子树角度来迭代改进它，从而使日光算法所得出的叶节点间的弧线相等，如图 4.2H 所示。日光算法首先在 PAUP*[17] 中实现。

分支图（cladogram）：分支图是指只显示树的拓扑结构而不设置枝长标尺的树图。如果想要绘制分支图，则只需要将 branch.length 参数的值设置为 "none"，对于所有的布局来说，我们都可以采用这种方法来绘制分支图。

时间尺度布局（time-scaled）：在绘制时间尺度树时，我们必须将最近采样日期信息传递给 mrsd 参数，ggtree() 函数会根据采样时间（叶节点）及分歧时间（内部节点）来设置树的标尺，并默认在树的下方生成一个时间坐标轴。用户可以使用 deeptime 包将地质年代信息（如宙、代、纪等）添加到 ggtree() 函数生成的图中。

```
beast_file <- system.file("examples/MCC_FluA_H3.tree",
                          package="ggtree")
beast_tree <- read.beast(beast_file)
ggtree(beast_tree, mrsd="2013-01-01") + theme_tree2()
```

二维树布局（two-dimensional）：二维树是指系统发育树在由相关联表型信息（数值型或分类型性状数据均可作为 y 轴）及枝长信息（进化距离、分歧时间等均可作为 x 轴）所构建的空间中的投影。其中，表型信息可以用于衡量分类单元与其假定的祖先在某些生物学特征上的差异；还可以用于追踪随着病毒进化（x

轴）而发生改变的病毒表型，或者其他的一些病毒行为（y 轴）。而实际上，对于病毒表型或基因型随进化时间变化的分析已经被广泛应用于流感病毒进化的研究中 [18]，但这些研究中用到的分析图并不像二维树布局这样将各个数据点用对应的分支连接起来。因此，相比之下，我们通过二维树布局会更加方便分析这类数据，同时在大型序列数据集中也能更好地呈现结果。

  本示例使用了前文提到的 H3 人猪流感病毒的时间尺度树（见图 4.4，数据发表在本章参考文献 [19] 中），并根据预测的血凝素蛋白上的 N- 连接糖基化（NLG）位点数量来对各个分类单元及祖先序列的 y 轴位置进行调整，其中的 NLG 位点是使用 NetNGlyc 1.0 Server 预测得出的。我们可以通过将 ggtree() 函数中的 yscale 参数设置为一个数值变量或分类变量，以此来对 y 轴标尺进行调整。如果像本示例中一样，yscale 参数作为分类变量，则还需要通过 yscale_mapping 参数指定如何将分类变量映射到数值变量。

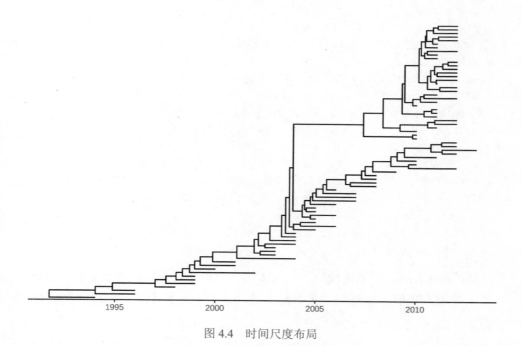

图 4.4 时间尺度布局

  x 轴为时间尺度（以年为单位）。本示例中的分歧时间是由 BEAST 软件使用分子钟模型推断出来的。

```
NAG_file <- system.file("examples/NAG_inHA1.txt", package="ggtree")
```

```
NAG.df <- read.table(NAG_file, sep="\t", header=FALSE,
                     stringsAsFactors = FALSE)

NAG <- NAG.df[,2]
names(NAG) <- NAG.df[,1]

## 根据宿主物种类型分割树
tip <- as.phylo(beast_tree)$tip.label
beast_tree <- groupOTU(beast_tree, gtip[grep("Swine", tip)],
                       group_name = "host")

p <- ggtree(beast_tree, aes(color=host), mrsd="2013-01-01",
            yscale = "label", yscale_mapping = NAG) +
  theme_classic() + theme(legend.position='none') +
  scale_color_manual(values=c("blue", "red"),
                     labels=c("human", "swine")) +
  ylab("Number of predicted N-linked glycoslyation sites")

## （可选）添加更多注释层以帮助读者更好地理解图
p + geom_nodepoint(color="grey", size=3, alpha=.8) +
  geom_rootpoint(color="black", size=3) +
  geom_tippoint(size=3, alpha=.5) +
  annotate("point", 1992, 5.6, size=3, color="black") +
  annotate("point", 1992, 5.4, size=3, color="grey") +
  annotate("point", 1991.6, 5.2, size=3, color="blue") +
  annotate("point", 1992, 5.2, size=3, color="red") +
  annotate("text", 1992.3, 5.6, hjust=0, size=4, label="Root node") +
  annotate("text", 1992.3, 5.4, hjust=0, size=4,
           label="Internal nodes") +
  annotate("text", 1992.3, 5.2, hjust=0, size=4,
           label="Tip nodes (blue: human; red: swine)")
```

如图 4.5 所示，二维树很适合呈现系统发育树中表型随着进化变化的过程。在本示例中，我们可以看到人类甲型流感病毒的 H3 基因在过去 20 年始终保持着高水平的 N- 连接糖基化位点数量（n 保持范围为 8～9），而有一个谱系的病毒中的 NLG 位点数却骤降至 5～6，开始传播到猪群，并以此为基础继续发展。也确实有人曾提出过类似的假说，假定在病毒血凝素蛋白上存在高度糖基化的人流感病毒为保护抗原位点不受群体免疫的影响提供了更好的保护作用，因此相比于正在流行的人流感病毒株，在对人流感病毒有着较强群体免疫的人类群体中存在着选择优势。而对于这条跨越了种间屏障并传播到猪群的谱系来说,由高水平表面聚糖提供的屏蔽作用反而加大了其选择劣势,因为其受体结合域可能也会被这种作用屏蔽,从而极大地影响了此谱系的病毒对新宿主物种的适应性。

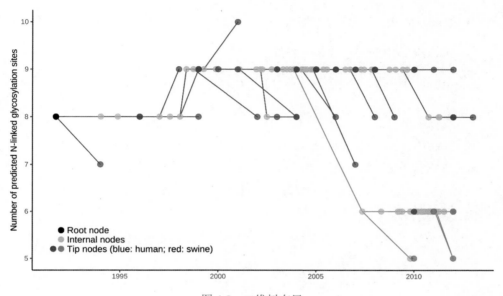

图4.5　二维树布局

树的树干及其他分支分别以红色(猪)及蓝色(人类)突出显示。x轴以时间尺度树的枝长(以年为单位)为标尺,y轴以节点属性变量为标尺。本示例为预测的血凝素蛋白上的N-连接糖基化位点数量。不同类型的节点以不同颜色的圆圈形式呈现。需要注意的是,拥有相同的x轴、y轴坐标(本示例中分别为时间及NLG位点数量)的节点将会被堆叠为一个点,并通过叠加该位置上所有节点的颜色来对其进行着色。

4.3　绘制树的构成部分

4.3.1　绘制树的标尺

用户可以使用geom_treescale()图层函数来添加树的标尺,如图4.6A～图4.6C所示。

```
ggtree(tree) + geom_treescale()
```

geom_treescale() 图层函数支持以下参数。

- x 和 y：用于调整标尺的位置。
- width：用于调整标尺的长度。
- fontsize：用于调整文本的大小。
- lineszie：用于调整线条的粗细。
- offset：用于调整线条与文本之间的距离。
- color：用于调整标尺的颜色。

```
ggtree(tree) + geom_treescale(x=0, y=45, width=1, color='red')
ggtree(tree) + geom_treescale(fontsize=6, linesize=2, offset=1)
```

我们还可以使用 theme_tree2() 函数，以添加 x 轴的形式来绘制树的标尺，如图 4.6D 所示。

```
ggtree(tree) + theme_tree2()
```

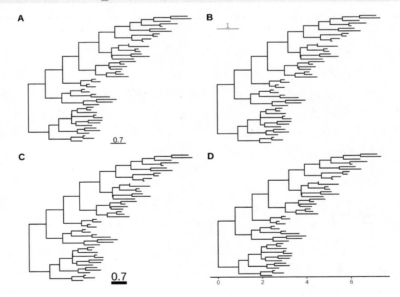

图 4.6 绘制树的标尺

使用 geom_treescale() 图层函数自动添加表示进化距离的标尺（A）。用户可以修改标尺的颜色、长度和位置（B）。也支持修改标尺上的文本字号，以及文本与标尺间的相对位置（C）。也可以通过添加 x 轴的形式添加标尺（D），这种方法比较适合绘制时间尺度树。

当然，树并不局限于以进化距离作为标尺。treeio 还可以依据其他的数值变量来对树的比例进行重新调整；ggtree 也支持用户在可视化时手动指定一个数值变量来作为树的枝长。

4.3.2 绘制内/外部节点

我们可以使用 geom_nodepoint() 图层函数、geom_tippoint() 图层函数或 geom_point() 图层函数在树的内/外部节点处添加点图层。

```
ggtree(tree) +
geom_point(aes(shape=isTip, color=isTip), size=3)

p <- ggtree(tree) +
geom_nodepoint(color="#b5e521", alpha=1/4, size=10)
p + geom_tippoint(color="#FDAC4F", shape=8, size=3)
```

图 4.7 所示为绘制内/外部节点。

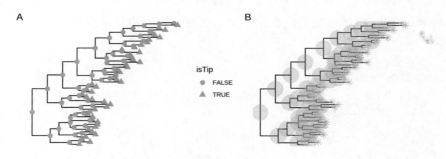

图 4.7　绘制内/外部节点

使用 geom_point() 图层函数自动为节点添加符号点（A）。使用 geom_nodepoint() 图层函数只为内部节点添加符号点，使用 geom_tippoint() 图层函数只为外部节点添加符号点（B）。

4.3.3 绘制标签

用户可以使用 geom_text() 图层函数或 geom_label() 图层函数为所有节点添加标签（如果存在对应节点标签信息），或者使用 geom_tiplab() 图层函数只为叶节点添加标签，如图 4.8A 所示。

```
p + geom_tiplab(size=3, color="purple")
```

geom_tiplab() 图层函数除了支持通过 text 或 label 几何对象对标签进行可视化，还支持通过 image 几何对象使用图像文件为叶节点添加标签（参见第 7 章）。ggtree 包还提供了与 geom_tiplab() 图层函数对应的 geom_nodelab() 图层函数来对内部节点标签进行单独绘制。

对于环形布局与无根布局来说，ggtree 包还支持根据分支的角度来旋转节点标签，如图 4.8B 所示。

```
ggtree(tree) + geom_tiplab(as_ylab=TRUE, color='firebrick')
```

当遇到叶节点标签过长的情况时，标签可能会被截断，其解决方法是，将叶节点标签显示为 y 轴标签，如图 4.8C 所示。但是这种方法只适用于矩形布局和树状图布局，同时在使用该方法时，用户需要通过 theme() 函数来调整叶节点标签。在附录 A 常见问题里还给出了另一种解决方法。

```
ggtree(tree) + geom_tiplab(as_ylab=TRUE, color='firebrick')
ggtree(tree) + geom_tiplab(as_ylab=TRUE, color='firebrick')
```

在默认情况下，这些文本标签会被绘制于节点的位置。我们也可以通过设置 aes(x = branch) 将其绘制于分支的中央，这点在注释由父节点到子节点的过渡时非常实用。

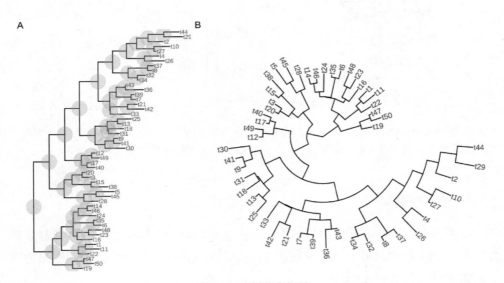

图 4.8　绘制节点标签

第 4 章　系统发育树可视化

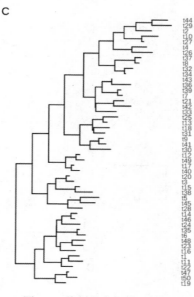

图 4.8　绘制节点标签（续）

我们可以通过 geom_tiplab() 图层函数来绘制叶节点标签（A）。在环形布局、扇形布局或无根布局中，标签会随着分支的角度自动旋转（B）。在矩形布局或树状图布局中，叶节点标签也可以以 y 轴标签的方式呈现（C）。

4.3.4　绘制根分支

在默认情况下，ggtree() 函数并不将根分支绘制出来。用户可以使用 geom_rootedge() 图层函数自动绘制根分支，如图 4.9A 所示。如果没有找到根分支相关的信息，则不会绘制任何内容，如图 4.9B 所示。在这种情况下，用户可以在树中设置根分支的值，如图 4.9C 所示，或者在 geom_rootedge() 图层函数中指定 rootedge 参数的值，如图 4.9D 所示。通过为树设置较长的根分支，可以大大提高环形布局的树的可读性。

```
## 将根分支长度设置为 1 时
tree1 <- read.tree(text='((A:1,B:2):3,C:2):1;')
ggtree(tree1) + geom_tiplab() + geom_rootedge()

## 当没有根分支信息时
tree2 <- read.tree(text='((A:1,B:2):3,C:2);')
ggtree(tree2) + geom_tiplab() + geom_rootedge()
```

```
## 手动设置根分支长度
tree2$root.edge <- 2
ggtree(tree2) + geom_tiplab() + geom_rootedge()

## 仅在可视化阶段指定根分支长度
## 这样做将忽略原本的 tree2$root.edge 值
ggtree(tree2) + geom_tiplab() + geom_rootedge(rootedge = 3)
```

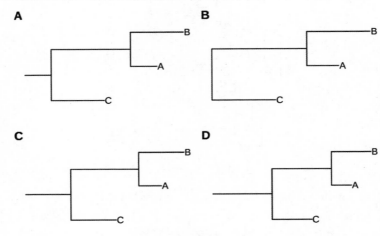

图 4.9　绘制根分支

geom_rootedge() 图层函数用于绘制进化树的根分支（A）。如果没有根分支信息，则不会绘制任何内容（B）。在这种情况下，用户可以为树添加根分支信息（C），或者仅在可视化阶段指定根分支的长度（D）。

4.3.5　给树着色

有了 ggtree[6] 包，我们可以非常轻松地实现对系统发育树的着色，只需要通过 aes(color=VAR) 命令将指定的特征（连续型或离散型均可，见图 4.10）映射到树的颜色属性。

```
ggtree(beast_tree, aes(color=rate)) +
    scale_color_continuous(low='darkgreen', high='red') +
    theme(legend.position="right")
```

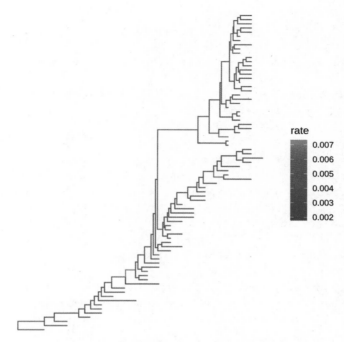

图 4.10 通过连续型特征或离散型特征给树着色

本示例分支依据对应子节点的值进行着色。

用户可以使用任何可用的特征来作为树颜色的标尺，如进化枝后验概率或 dN/dS 等。ggtree 包还为连续型特征提供了 continuous 参数。通过设置该参数，我们可以在分支上绘制出连续状态转换的效果。为此，我们准备了一个示例树[1]来演示这个功能，如图 4.11A 所示。如果想要为树添加一个黑色边框，则可以在树的下方绘制一棵纯黑且分支略粗一点的树，就可以达到这样的效果，如图 4.11B 所示。

```
library(ggtree )
library(treeio)
library(tidytree)
library(ggplot2)
library(TDbook)
## 参考 http://www.phytools.org/eqg2015/asr.html
##
## 从 'TDbook' 加载 'tree_anole' and 'df_svl'
svl <- as.matrix(df_svl)[,1]
fit <- phytools::fastAnc(tree_anole, svl, vars=TRUE, CI=TRUE)
```

[1] 示例来源于 phytools 网站，详情请参见"外链资源"文档中第 4 章第 1 条

```r
td <- data.frame(node = nodeid(tree_anole, names(svl)),
                 trait = svl)
nd <- data.frame(node = names(fit$ace), trait = fit$ace)

d <- rbind(td, nd)
d$node <- as.numeric(d$node)
tree <- full_join(tree_anole, d, by = 'node')

p1 <- ggtree(tree, aes(color=trait), layout = 'circular',
        ladderize = FALSE, continuous = TRUE, size=2) +
    scale_color_gradientn(colours=c("red", 'orange', 'green', 'cyan', 'blue')) +
    geom_tiplab(hjust = -.1) +
    xlim(0, 1.2) +
    theme(legend.position = c(.05, .85))

p2 <- ggtree(tree, layout='circular', ladderize = FALSE, size=2.8) +
    geom_tree(aes(color=trait), continuous=T, size=2) +
    scale_color_gradientn(colours=c("red", 'orange', 'green', 'cyan', 'blue')) +
    geom_tiplab(aes(color=trait), hjust = -.1) +
    xlim(0, 1.2) +
    theme(legend.position = c(.05, .85))

plot_list(p1, p2, ncol=2, tag_levels="A")
```

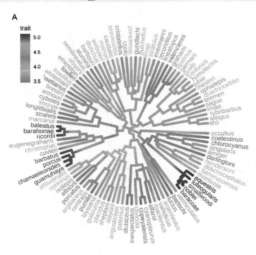

图 4.11 在分支上绘制出连续状态转换的效果

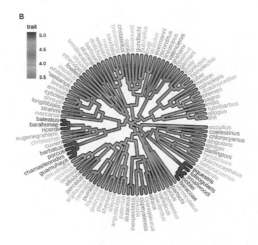

图 4.11　在分支上绘制出连续状态转换的效果（续）

本示例分支依据祖先性状至后代性状的值来着色。

除此之外，我们还可以以二维树的形式，以表型信息为 y 轴标尺来构建表型图（phenogram，见图 4.12）。当叶节点标签相互堆叠时，我们可以借助 ggrepel 包使叶节点标签相互排斥，从而避免叶节点标签相互堆叠。

```
ggtree(tree, aes(color=trait), continuous = TRUE, yscale = "trait") +
  scale_color_viridis_c() + theme_minimal()
```

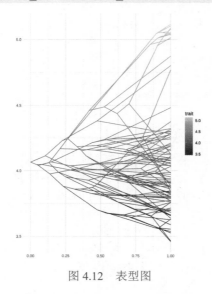

图 4.12　表型图

将进化树投影到以时间信息（或遗传距离）为 x 轴，表型信息为 y 轴构建的空间中。

4.3.6 调整进化树标尺

系统发育树通常都是以进化距离（平均每个位点的替换数）作为标尺的。通过 ggtree 包，用户可以指定任何由进化分析得出的数值变量（如 dN/dS），并以此重新调整进化树的标尺。

在本示例中，我们绘制了一个时间尺度树，如图 4.13A 所示，并以 BEAST 推断的位点替换速率作为标尺对其分支比例进行了调整，如图 4.13B 所示。

```
library("treeio")
beast_file <- system.file("examples/MCC_FluA_H3.tree", package="ggtree")
beast_tree <- read.beast(beast_file)
beast_tree
```

```
## 'treedata' S4 object that stored information
## of
##    '/home/ygc/R/library/ggtree/examples/MCC_FluA_H3.tree'.
##
## ...@ phylo:
##
## Phylogenetic tree with 76 tips and 75 internal nodes.
##
## Tip labels:
##    A/Hokkaido/30-1-a/2013, A/New_York/334/2004,
## A/New_York/463/2005, A/New_York/452/1999,
## A/New_York/238/2005, A/New_York/523/1998, ...
##
## Rooted; includes branch lengths.
##
## with the following features available:
##    'height', 'height_0.95_HPD', 'height_median',
## 'height_range', 'length', 'length_0.95_HPD',
## 'length_median', 'length_range', 'posterior', 'rate',
## 'rate_0.95_HPD', 'rate_median', 'rate_range'.
```

```
p1 <- ggtree(beast_tree, mrsd='2013-01-01') + theme_tree2() +
    labs(caption="Divergence time")
p2 <- ggtree(beast_tree, branch.length='rate') + theme_tree2() +
    labs(caption="Substitution rate")
```

在另一个示例中，我们绘制了一个由 CODEML 推断而得的树（见图 4.13C），并以 dN/dS 值作为标尺对其分支的比例进行了调整（见图 4.13D）。

```
mlcfile <- system.file("extdata/PAML_Codeml", "mlc", package="treeio")
mlc_tree <- read.codeml_mlc(mlcfile)
p3 <- ggtree(mlc_tree) + theme_tree2() +
    labs(caption="nucleotide substitutions per codon")
p4 <- ggtree(mlc_tree, branch.length='dN_vs_dS') + theme_tree2() +
    labs(caption="dN/dS tree")
```

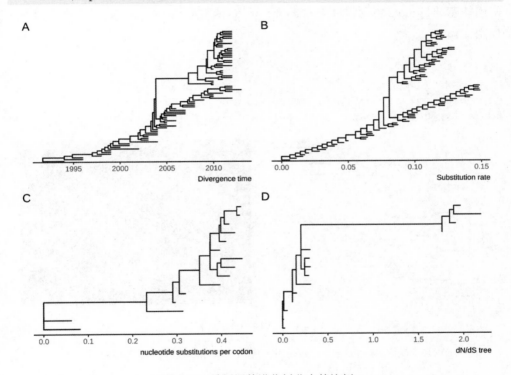

图 4.13 重新调整进化树分支的比例

由 BEAST 推断的时间尺度树（A）。以位点替换速率标尺重新调整其分支的比例后的结果（B）。由 CODEML 推断的树（C），以及以 dN/dS 值作为标尺重新调整其分支比例后的结果（D）。

这个功能为我们通过可视化的方式探索进化树关联数据及树结构之间的关系带来了极大的便利。除了在可视化阶段通过指定 branch.length 参数来改变枝长比例，用户还可以通过 treeio 包[15]中的 rescale_tree() 函数直接更改存储在树对象中的枝长信息。我们可以使用 rescale_tree() 函数生成与图 4.13B 相同的树。

```
beast_tree2 <- rescale_tree(beast_tree, branch.length='rate')
ggtree(beast_tree2) + theme_tree2()
```

4.3.7 修改主题组件

我们使用 ggtree 包中的 theme_tree() 函数可以在一个完全空白的画布上绘制进化树，而 theme_tree2() 函数则在此基础上通过添加 x 轴增加了对系统发育距离的呈现。这两个函数都可以接收用于定义背景颜色的 bgcolor 参数。我们也可以使用任意的 ggplot2 主题组件来对 theme_tree() 函数或 theme_tree2() 函数进行修改，如图 4.14 所示。

```
set.seed(2019)
x <- rtree(30)
ggtree(x, color="#0808E5", size=1) + theme_tree("#FEE4E9")
ggtree(x, color="orange", size=1) + theme_tree('grey30')
```

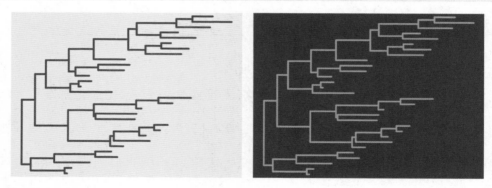

图 4.14　两种不同的主题

我们可以在 theme_tree() 函数或 theme_tree2() 函数中添加或修改任意的 ggplot2 主题组件。

ggtree 包还支持使用图像文件作为进化树的背景，详情请参考附录 B 中的示例。

4.4　对树列表进行可视化

ggtree 包提供了对 multiPhylo 对象及 treedataList 对象的支持，可以同时对列表中的所有树进行可视化。这些树可以以相互堆叠的方式呈现，也可以通过 facet_wrap() 函数或 facet_grid() 函数绘制于不同的面板中，如图 4.15 所示。

```
## trees <- lapply(c(10, 20, 40), rtree)
## class(trees) <- "multiPhylo"
## ggtree(trees) + facet_wrap(~.id, scale="free") + geom_tiplab()

f <- system.file("extdata/r8s", "H3_r8s_output.log", package="treeio")
r8s <- read.r8s(f)
ggtree(r8s) + facet_wrap( ~.id, scale="free") + theme_tree2()
```

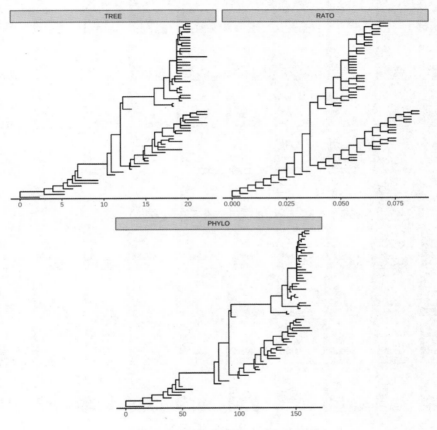

图 4.15　对 multiPhylo 对象进行可视化

ggtree 包支持同时对存储在 multiPhylo 对象或 treedataList 对象中的多个树进行可视化。

我们可以做到同时对 100 棵树进行可视化，如图 4.16 所示。这使得科研工作者能同时探索大量的系统发育树，并找出其中的一致树或异构树，而其中的一致树也可以以密度树的形式来呈现。

```
btrees <- read.tree(system.file("extdata/RAxML",
                                "RAxML_bootstrap.H3",
```

```
                              package="treeio")
                    )
ggtree(btrees) + facet_wrap(~.id, ncol=10)
```

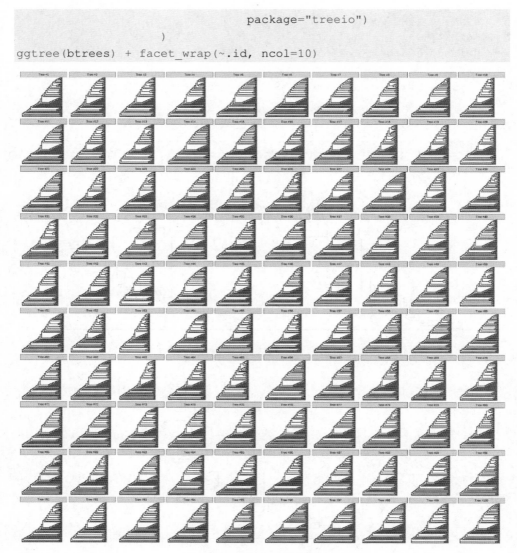

图 4.16 同时对 100 棵树进行可视化

4.4.1 使用不同变量的值注释同一棵树

如果想要对同一棵树以不同变量的值呈现出不同的可视化的效果，则可以把它们先分别绘制出来，再通过 patchwork 或 aplot 将它们并排组合起来。

还有一种方法是利用 ggtree 绘制树列表的功能，我们可以通过 ggtree 中的

subset 美学映射属性或 td_filter() 函数，在特定面板中筛选出特定的变量，并使用它对进化树进行注释，如图 4.17 所示。示例代码中的 ".id" 用于存储不同树 ID 的保留变量。

```
set.seed(2020)
x <- rtree(30)
d <- data.frame(label=x$tip.label, var1=abs(rnorm(30)),
var2=abs(rnorm(30)))
tree <- full_join(x, d, by='label')
trs <- list(TREE1 = tree, TREE2 = tree)
class(trs) <- 'treedataList'
ggtree(trs) + facet_wrap(~.id) +
  geom_tippoint(aes(subset=.id == 'TREE1', colour=var1)) +
  scale_colour_gradient(low='blue', high='red') +
  ggnewscale::new_scale_colour()    +
  geom_tippoint(aes(colour=var2), data=td_filter(.id == "TREE2")) +
  scale_colour_viridis_c()
```

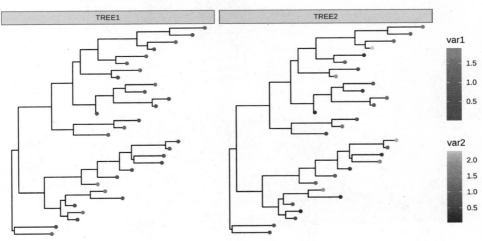

图 4.17　使用不同变量值注释同一棵树

使用 subset 美学映射属性（TREE1 面板）或 td_filter() 函数（TREE2 面板）对特定面板中的特定变量进行筛选。

4.4.2　密度树

我们还可以通过 ggdensitree() 函数将多个自举树融合起来，形成密度树，如图 4.18 所示。这种形式可以帮助我们发现大量的树之间的共性与差异。这些树会被堆叠在一起，树的结构也会发生旋转，以确保叶节点顺序保持一致。叶节点的顺序可由 tip.order 参数指定，在默认情况下（tip.order = 'mode'），叶节点会以出现频率最高的拓扑结构排列。我们也可以通过传入一个字符向量来指定叶节点顺序，或者传入一个整数 n 来指定使用第 n 棵树的叶节点排列顺序。如果将"mds"传入 tip.order 参数，则会对叶节点之间的路径距离进行多维尺度分析（Multidimensional Scaling，MDS），并依此对叶节点进行排序；或者将"mds_dist"传入 tip.order 参数，这样就会依据叶节点之间距离的 MDS 结果来对叶节点进行排序。

```
ggdensitree(btrees, alpha=.3, colour='steelblue') +
    geom_tiplab(size=3) + hexpand(.35)
```

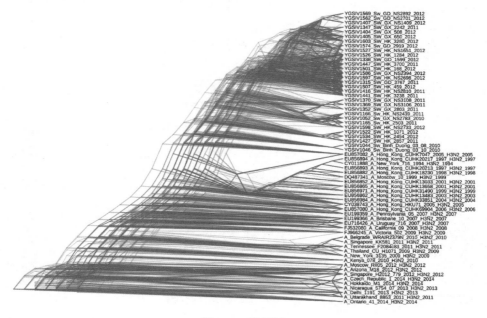

图 4.18　密度树

由大量的树相互堆叠而成，同时为了保证叶节点顺序的一致性不被破坏，树的结构也依此发生了旋转。

4.5　总结

我们通过 ggtree 包可以轻松地对系统发育树进行可视化，只需要通过

ggtree(tree) 一行命令即可完成。ggtree 包还提供了多种几何图层以便于展示进化树的构成部分，如叶节点标签、内/外部节点均可添加的符号点、根分支等。进化树关联数据可以用于重新调整枝长标尺、给进化树着色或者直接展示在树上。这一切都要归功于 ggplot2 的图形语法，它的存在让添加多个图层并对树图进行自定义的过程变得非常轻松（通过 ggplot2 中的主题及标尺功能）。ggtree 包还提供了一些专门为进化树的注释设计的图层，让进化树及其关联数据的呈现变得非常简单，简单树图能够快速生成，而同时复杂的树图可以通过图层的叠加轻松地生成。

4.6 本章练习题

1. 随机生成一棵树，并将其绘制出来。要求其为圆角矩形布局，向右上方倾斜 45 度。并在水平方向绘制叶节点标签（需要安装 ggplotify 包）。

2. 随机生成一棵树，使用 phytools 包中的 fastBM() 函数及 fastAnc() 函数随机生成连续型性状，在树上呈现出连续状态转化的效果，并在树的外部添加黑色边框。

3. 绘制一个简单的烟花。首先生成一棵随机树，以环形布局分支图（不设置枝长）的形式对其进行可视化，将树的颜色设置为透明；其次绘制内部节点，将颜色设置为红色，透明度设置为 0.5；然后绘制外部节点，将颜色设置为黄色，形状设置为"米"字型（shape = 8）；最后将背景颜色设置为黑色。

参考文献

[1] Page R D. Visualizing phylogenetic trees using TreeView[J]. Current Protocols in Bioinformatics, 2003(1): 6.2.1-6.2.15.

[2] Chevenet F, Brun C, Banuls A L, et al. TreeDyn: towards dynamic graphics and annotations for analyses of trees[J]. BMC Bioinformatics, 2006, 7: 439.

[3] Huson D H, Scornavacca C. Dendroscope 3: an interactive tool for rooted phylogenetic trees and networks[J]. Syst Biol, 2012, 61(6): 1061-1067.

[4] He Z, Zhang H, Gao S, et al. Evolview v2: an online visualization and management tool

for customized and annotated phylogenetic trees[J]. Nucleic Acids Res, 2016, 44(W1): W236-W241.

[5] Letunic I, Bork P. Interactive Tree Of Life (iTOL): an online tool for phylogenetic tree display and annotation[J]. Bioinformatics, 2007, 23(1): 127-128.

[6] Yu G, Smith D K, Zhu H, et al. ggtree: an R package for visualization and annotation of phylogenetic trees with their covariates and other associated data[J]. Methods in Ecology and Evolution, 2016, 8(1): 28-36.

[7] Gentleman R C, Carey V J, Bates D M, et al. Bioconductor: open software development for computational biology and bioinformatics[J]. Genome Biol, 2004, 5(10): R80.

[8] Team R C. R: A language and environment for statistical computing. R Foundation for Statistical Computing[EB/OL]. https://www.R-project.org/.

[9] Wilkinson L W D R. The Grammar of Graphics [M]. 2nd edition. Springer, 2005.

[10] Wickham H. ggplot2: Elegant graphics for data analysis[M]. 2016.

[11] Paradis E, Claude J, Strimmer K. APE: analyses of phylogenetics and evolution in R language[J]. Bioinformatics, 2004, 20(2): 289-290.

[12] Revell L J. phytools: an R package for phylogenetic comparative biology (and other things)[J]. Methods in Ecology and Evolution, 2012, 3(2): 217-223.

[13] Jombart T, Aanensen D M, Baguelin M, et al. OutbreakTools: a new platform for disease outbreak analysis using the R software[J]. Epidemics, 2014, 7: 28-34.

[14] McMurdie P J H S. phyloseq: an R package for reproducible interactive analysis and graphics of microbiome census data[J]. Plos One, 2013, 8(4):e61217.

[15] Wang L, Lam T T, Xu S, et al. treeio: an R package for phylogenetic tree input and output with richly annotated and associated data[J]. Molecular Biology and Evolution, 2020, 37(2): 599-603.

[16] Retief J D. Phylogenetic analysis using PHYLIP[J]. Methods Mol Biol, 2000, 132: 243-258.

[17] Wilgenbusch J C, Swofford D. Inferring evolutionary trees with PAUP*[J]. Current Protocols in Bioinformatics, 2003(1): 6.4.1-6.4.28.

[18] Neher R A, Bedford T, Daniels R S, et al. Prediction, dynamics, and visualization of antigenic phenotypes of seasonal influenza viruses[J]. Proc Natl Acad Sci U S A, 2016, 113(12): E1701-E1709.

[19] Liang H, Lam T T, Fan X, et al. Expansion of genotypic diversity and establishment of 2009 H1N1 pandemic-origin internal genes in pigs in China[J]. J Virol, 2014, 88(18): 10864-10874.

第 5 章 系统发育树注释

5.1 使用图形语法对树进行可视化及注释

ggtree 包[1]被设计为能同时满足更为通用的或特定类型的进化树可视化及注释的需求。它支持 ggplot2 的图形语法,我们可以通过组合多个注释图层来随意地对进化树进行可视化或注释。如图 5.1 所示,我们使用了图形语法,通过"+"运算符组合多种注释图层来对 NHX 树进行注释。我们将物种信息标注在分支中间,复制事件及进化枝的自举值则被标注在最近共同祖先节点上。

```
library(ggtree)
treetext = "(((ADH2:0.1[&&NHX:S=human], ADH1:0.11[&&NHX:S=human]):
0.05 [&&NHX:S=primates:D=Y:B=100],ADHY:
0.1[&&NHX:S=nematode],ADHX:0.12 [&&NHX:S=insect]):
0.1[&&NHX:S=metazoa:D=N],(ADH4:0.09[&&NHX:S=yeast],
ADH3:0.13[&&NHX:S=yeast], ADH2:0.12[&&NHX:S=yeast],
ADH1:0.11[&&NHX:S=yeast]):0.1[&&NHX:S=Fungi])[&&NHX:D=N];"
tree <- read.nhx(textConnection(treetext))
ggtree(tree) + geom_tiplab() +
  geom_label(aes(x=branch, label=S), fill='lightgreen') +
  geom_label(aes(label=D), fill='steelblue') +
  geom_text(aes(label=B), hjust=-.5)
```

在本示例中,存储于 NHX 标签中的注释信息通过多个注释图层被展示在进化树上。例如,使用 geom_tiplab() 图层函数来展示叶节点标签(此示例中为基因名);使用 geom_label() 图层函数,以绿色标签的形式来展示物种信息(存储于 NHX 文件的 S 标签中),或者以钢青色标签来展示复制事件(存储于 NHX 文件的 D 标签中);使用 geom_text() 图层函数来展示进化枝的自举值(存储于 NHX 文件的 B 标签中)。

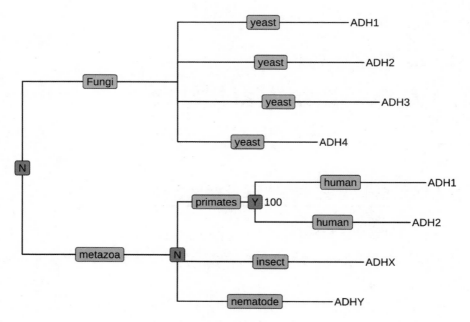

图 5.1　使用图形语法注释进化树

ggplot2 包中定义的图层可以直接运用于 ggtree 包中。如图 5.1 所示，我们可以使用 geom_label() 图层函数和 geom_text() 图层函数对进化树进行注释。但是 ggplot2 包中并没有提供专门为进化树注释设计的图层，如叶节点标签图层、枝长的标尺图例图层、进化枝的标签及突出显示图层等。为了使对进化树的注释变得更加方便灵活，ggtree 包实现了多种专门为进化树注释设计的图层，如表 5.1 所示，能更好地帮助用户对进化树的不同部分或组成部分进行多种不同方式的注释。

表 5.1　ggtree 中定义的图层函数

图层函数名称	说　　明
geom_balance()	突出显示一个内部节点的两个直系子进化枝
geom_cladelab()	使用条带及文本标签（或图片）注释进化枝
geom_facet()	将关联数据绘制于特定的面板，并将图与树对齐
geom_hilight()	使用环形布局或矩形布局突出显示选定的进化枝
geom_inset()	在节点处添加子图
geom_label2()	geom_label() 图层函数的改进版本，支持通过 subset 美学映射属性对数据进行筛选
geom_nodepoint()	使用符号点注释内部节点

续表

图层函数名称	说明
geom_point2()	geom_point() 图层函数的改进版本,支持通过 subset 美学映射属性对数据进行筛选
geom_range()	通过添加条带图层来呈现进化推论的不确定性
geom_rootpoint()	使用符号点注释根节点
geom_rootedge()	为树添加根分支
geom_segment2()	geom_segment() 图层函数的改进版本,支持通过 subset 美学映射属性对数据进行筛选
geom_strip()	使用线条及文本标签注释相关联的分类单元
geom_taxalink()	连接相关的分类单元
geom_text2()	geom_text() 图层函数的改进版本,支持通过 subset 美学映射属性对数据进行筛选
geom_tiplab()	添加叶节点标签图层
geom_tippoint()	使用符号点注释外部节点
geom_tree()	树结构图层,支持以多种布局呈现
geom_treescale()	添加枝长标尺图例

5.2 进化树注释图层

5.2.1 彩色条带

ggtree 包[1]实现了 geom_cladelab() 图层函数。用户可以通过该图层函数在指定的进化枝外部添加彩色条带及标签来对进化枝进行注释。

在使用 geom_cladelab() 图层函数时,需要输入一个内部节点的编号,将自动为节点对应的进化枝添加标签,如图 5.2A 所示。

```
set.seed(2015-12-21)
tree <- rtree(30)
p <- ggtree(tree) + xlim(NA, 8)

p + geom_cladelab(node=45, label="test label") +
    geom_cladelab(node=34, label="another clade")
```

用户可以通过参数 align = TRUE 来将进化枝标签在竖直方向对齐,或者通过 offset 参数来调整条带与树之间的距离,以及通过 textcolor 参数、barcolor 参数来调整文本与条带的颜色等，如图 5.2B 所示。

```
p + geom_cladelab(node=45, label="test label", align=TRUE,
            offset = .2, textcolor='red', barcolor='red') +
    geom_cladelab(node=34, label="another clade", align=TRUE,
            offset = .2, textcolor='blue', barcolor='blue')
```

用户也可以通过 angle 参数来改变文本的倾斜角度，以及通过 offset.text 参数来调整文本到条带之间的距离。而文本与条带的大小则可分别通过 fontsize 参数与 barsize 参数来调节，如图 5.2C 所示。

```
p + geom_cladelab(node=45, label="test label", align=TRUE, angle=270,
            hjust='center', offset.text=.5, barsize=1.5, fontsize=8) +
    geom_cladelab(node=34, label="another clade", align=TRUE, angle=45)
```

用户还可以通过设置 geom = 'label' 为文本添加标签，并通过 fill 参数来设置标签的背景颜色，如图 5.2D 所示。

```
p + geom_cladelab(node=34, label="another clade", align=TRUE,
            geom='label', fill='lightblue')
```

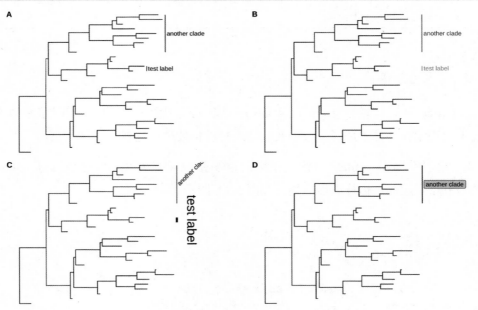

图 5.2 为进化枝添加标签

默认参数下的效果（A）；对条带和文本进行着色及对齐（B）。更改标签的大小和角度（C）。使用 geom_cladelab() 图层函数添加带有背景颜色的标签（D）。

除此之外，geom_cladelab() 图层函数也允许用户使用本地图片或 phylopic 来注释进化枝，还支持通过美学映射自动使用条带及文字标签或图像注释进化枝（如将变量映射到进化枝标签的颜色），如图 5.3 所示。

```r
dat <- data.frame(node = c(45, 34),
                  name = c("test label", "another clade"))
## 当 geom="text"、geom="label" 或 geom="shadowtext" 时,
## 需要指定 node 及 label 的映射
p1 <- p + geom_cladelab(data = dat,
         mapping = aes(node = node, label = name, color = name),
         fontsize = 3)
dt <- data.frame(node = c(45, 34),
                 image = c("7fb9bea8-e758-4986-afb2-95a2c3bf983d",
                           "0174801d-15a6-4668-bfe0-4c421fbe51e8"),
                 name = c("specie A", "specie B"))
# 当 geom="phylopic" 或 geom="image" 时, 需要指定 image 的映射
p2 <- p + geom_cladelab(data = dt,
              mapping = aes(node = node, label = name, image = image),
              geom = "phylopic", imagecolor = "black",
              offset=1, offset.text=0.5)
## 同样支持将变量映射至图像的颜色及大小
p3 <- p + geom_cladelab(data = dt,
              mapping = aes(node = node, label = name,
                    image = image, color = name),
              geom = "phylopic", offset = 1, offset.text=0.5)
```

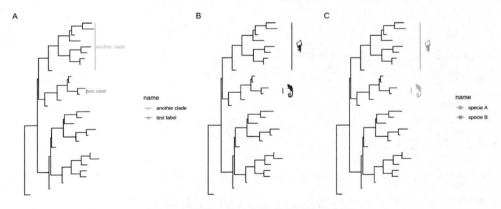

图 5.3　使用美学映射为进化枝添加标签

geom_cladelab() 图层函数允许用户使用美学映射来注释进化枝（A）；支持使用图像或 phylopic 来注释进化枝（B）；也支持将变量映射到标签的属性以改变图像、字号或颜色（C）。

geom_cladelab() 图层函数也支持无根布局，如图 5.4A 所示。

```
ggtree(tree, layout="daylight") +
  geom_cladelab(node=35, label="test label", angle=0,
                fontsize=8, offset=.5, vjust=.5) +
  geom_cladelab(node=55, label='another clade',
                angle=-95, hjust=.5, fontsize=8)
```

geom_cladelab() 是为标记单系群（进化枝）设计的图层函数。但有时相关联的分类单元并不能形成一个进化枝，为此我们在 ggtree 包中提供了另一个图层函数 geom_strip()。该图层函数可以为多系群或旁系群添加条带及可选的标签，来表明它们之间的关联，如图 5.4B 所示。

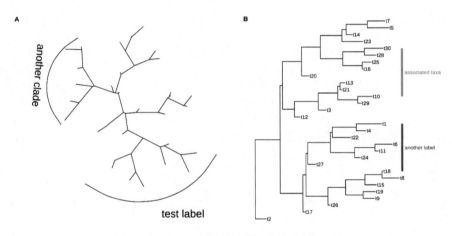

图 5.4　为相关分类单元添加标签

geom_cladelab() 是为添加单系群标签设计的图层函数，且支持无根布局（A）。而 geom_strip() 图层函数则可以为任意类型的关联分类单元添加标签，无论是单系群、多系群，还是旁系群（B）。

5.2.2　突出显示进化枝

ggtree 中的 geom_hilight() 图层函数用于接收一个内部节点号，并为选定的进化枝添加一个矩形图层以对其进行突出显示[①]，如图 5.5 所示。

```
nwk <- system.file("extdata", "sample.nwk", package="treeio")
tree <- read.tree(nwk)
```

---
① 如果想要在突出显示区域上绘制树，则请参见附录 A（A.7 在树的底部图层绘制图形），查看相关细节。

```
ggtree(tree) +
    geom_hilight(node=21, fill="steelblue", alpha=.6) +
    geom_hilight(node=17, fill="darkgreen", alpha=.6)

ggtree(tree, layout="circular") +
    geom_hilight(node=21, fill="steelblue", alpha=.6) +
    geom_hilight(node=23, fill="darkgreen", alpha=.6)
```

geom_hilight() 图层函数还支持以圆角形（将参数 type 设置为"encircle"）或矩形（将参数 type 设置为"rect"）来突出显示无根布局的树的进化枝，如图 5.5C 所示。

```
## 将参数 type 设置为 'encircle' 或 'rect'
pg + geom_hilight(node=55, linetype = 3) +
  geom_hilight(node=35, fill='darkgreen', type="rect")
```

为选定的进化枝设置不同颜色或线条类型也是一种将其突出显示的方法。

除了 geom_hilight() 图层函数，ggtree 包还实现了可以对指定的内部节点相邻的两个子进化枝进行突出显示的 geom_balance() 图层函数，如图 5.5D 所示。

```
ggtree(tree) +
  geom_balance(node=16, fill='steelblue', color='white', alpha=0.6, extend=1) +
  geom_balance(node=19, fill='darkgreen', color='white', alpha=0.6, extend=1)
```

geom_hilight() 图层函数支持使用美学映射来自动对进化枝进行突出显示，如图 5.5E、图 5.5F 所示，同时对于笛卡尔坐标系下的树图（如矩形布局的树），这些用于突出显示的矩形可以为圆角填充（见图 5.5E）或渐变色填充（见图 5.5F）。

```
## 使用外部数据
d <- data.frame(node=c(17, 21), type=c("A", "B"))
ggtree(tree) + geom_hilight(data=d, aes(node=node, fill=type),
                            type = "roundrect")
## 使用存储在树对象中的数据
x <- read.nhx(system.file("extdata/NHX/ADH.nhx", package="treeio"))
ggtree(x) + geom_hilight(mapping=aes(subset = node %in% c(10, 12),
                                    fill = S),
                         type = "gradient", graident.direction = 'rt',
                         alpha = .8) +
  scale_fill_manual(values=c("steelblue", "darkgreen"))
```

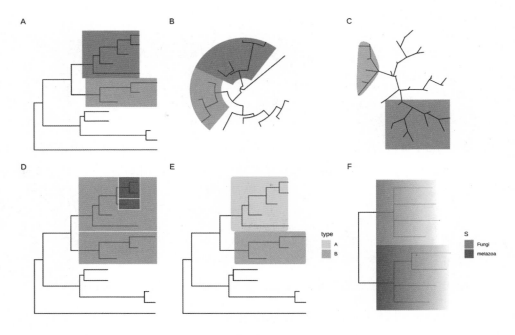

图 5.5　突出显示选定的进化枝

在矩形布局（A）、环形布局或扇形布局（B）及无根布局（C）下使用 grom_hilight() 图层函数，既突出显示指定的内部节点相邻的两个子进化枝（D），又突出显示使用树关联数据的进化枝（E 和 F）。

5.2.3　连接分类单元

有些进化事件（如基因重配或水平基因转移）无法通过简单的树直接建模，为此 ggtree 提供了 geom_taxalink() 图层函数。该图层函数可用于在树的任意两个节点之间绘制直线或曲线，也可用于通过连接分类单元的形式来呈现进化事件，并且该图层函数适用于矩形布局（见图 5.6A）、环形布局（见图 5.6B）和向内环形布局（见图 5.6C）。除此之外，geom_taxalink() 图层函数还可用于结合多个进化树，并以此呈现物种之间的关系或相互作用[2]。

geom_taxalink() 图层函数同样支持美学映射，如图 5.6D 所示，它需要输入一个数据框，其中存储着包含或不包含元数据的关联信息。

下面示例展示了 geom_taxalink() 图层函数的用法。

```r
p1 <- ggtree(tree) + geom_tiplab() + geom_taxalink(taxa1='A', 
taxa2='E') +
  geom_taxalink(taxa1='F', taxa2='K', color='red', linetype = 'dashed',
    arrow=arrow(length=unit(0.02, "npc")))

p2 <- ggtree(tree, layout="circular") +
     geom_taxalink(taxa1='A', taxa2='E', color="grey", alpha=0.5,
               offset=0.05, arrow=arrow(length=unit(0.01, "npc"))) +
     geom_taxalink(taxa1='F', taxa2='K', color='red',
               linetype = 'dashed', alpha=0.5, offset=0.05,
               arrow=arrow(length=unit(0.01, "npc"))) +
     geom_taxalink(taxa1="L", taxa2="M", color="blue", alpha=0.5,
               offset=0.05, hratio=0.8,
               arrow=arrow(length=unit(0.01, "npc"))) +
     geom_tiplab()

# 当树呈向内环形布局时，我们可以通过设置 outward = FALSE 来生成向内的曲线
p3 <- ggtree(tree, layout="inward_circular", xlim=150) +
     geom_taxalink(taxa1='A', taxa2='E', color="grey", alpha=0.5,
                   offset=-0.2, outward=FALSE,
                   arrow=arrow(length=unit(0.01, "npc"))) +
     geom_taxalink(taxa1='F', taxa2='K', color='red',
                   linetype = 'dashed',
                   alpha=0.5, offset=-0.2, outward=FALSE,
                   arrow=arrow(length=unit(0.01, "npc"))) +
     geom_taxalink(taxa1="L", taxa2="M", color="blue", alpha=0.5,
                   offset=-0.2, outward=FALSE,
                   arrow=arrow(length=unit(0.01, "npc"))) +
     geom_tiplab(hjust=1)

dat <- data.frame(from=c("A", "F", "L"),
                  to=c("E", "K", "M"),
                  h=c(1, 1, 0.1),
                  type=c("t1", "t2", "t3"),
                  s=c(2, 1, 2))
p4 <- ggtree(tree, layout="inward_circular", xlim=c(150, 0)) +
        geom_taxalink(data=dat,
                      mapping=aes(taxa1=from,
                                  taxa2=to,
                                  color=type,
                                  size=s),
                      ncp=10,
```

```
                        offset=0.15) +
          geom_tiplab(hjust=1) +
          scale_size_continuous(range=c(1,3))
plot_list(p1, p2, p3, p4, ncol=2, tag_levels='A')
```

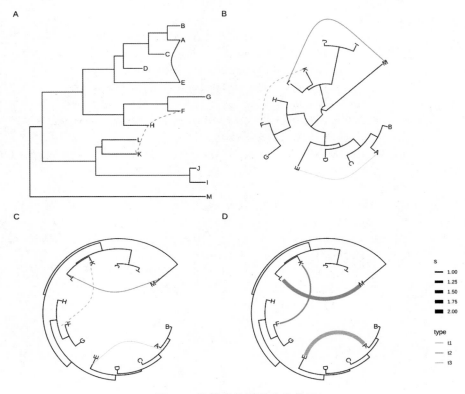

图 5.6 连接相关联的分类单元

我们可以以此来呈现进化事件或物种之间的关系。矩形布局（A）、环形布局（B）和向内环形布局（C 和 D）支持通过美学映射将变量映射至线条的粗细及颜色上（D）。

5.2.4 进化推论的不确定性

geom_range() 图层函数支持将区间（如最大后验密度、置信区间、极差等）以树节点上的水平条带呈现，而所对应的节点则会作为区间的中心点。在默认情况下，条带的中心代表了区间的平均值，如图 5.7A 所示。我们也可以通过设置 center 参数将中心点改为估算的均值、中值或观测值，如图 5.7B 所示。由于分支和区间可能并不在同一个尺度上，ggtree 包也提供了 scale_x_range() 函数来为区

间添加第二个 *x* 轴,如图 5.7C 所示。需要注意的是,默认的主题会将 *x* 轴隐藏起来,如果想要显示 *x* 轴,则可以使用 theme_tree2() 函数来实现。

```
file <- system.file("extdata/MEGA7", "mtCDNA_timetree.nex",
                    package = "treeio")
x <- read.mega(file)
p1 <- ggtree(x) + geom_range('reltime_0.95_CI', color='red', size=3,
alpha=.3)
p2 <- ggtree(x) + geom_range('reltime_0.95_CI', color='red', size=3,
                             alpha=.3, center='reltime')
p3 <- p2 + scale_x_range() + theme_tree2()
```

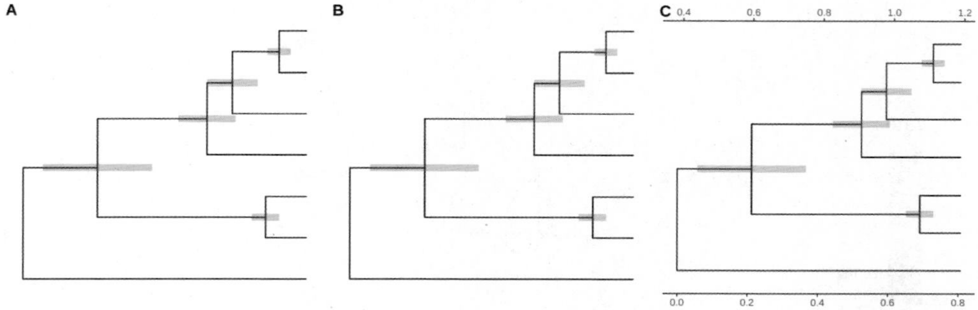

图 5.7　在进化树上展示进化推论的不确定性

区间的中心[区间的平均值(A)或估计值的平均值(B)]将被锚定于对应的节点。为区间添加第二个 *x* 轴以作为区间的标尺(C)。

5.3　使用进化软件输出结果注释树

我们利用 treeio 包[3] 可以解析不同格式的树文件及常用的软件输出结果,以此获取系统发育相关数据,并将其以 S4 对象的形式导入 R 中。然后,我们可以直接使用 ggtree 包对这些数据进行可视化。图 5.1 演示了如何使用 NHX 文件中存储的信息(物种分类信息、复制事件及自举值)对树进行注释。通过 PHYLDOG 或 RevBayes 输出的 NHX 文件可以由 treeio 包进行解析,并通过 ggtree 包使用其中的推论数据对树进行注释。

除此之外,BEAST、MrBayes 和 RevBayes 推断的进化数据,CODEML 推断的 dN/dS 值,HyPhy、CODEML 或 BASEML 推断的祖先序列,以及 EPA 和

PPLACER 推断的短读序列放置信息均可用来直接对树进行注释。

```
file <- system.file("extdata/BEAST", "beast_mcc.tree", package="treeio")
beast <- read.beast(file)
ggtree(beast, aes(color=rate)) +
    geom_range(range='length_0.95_HPD', color='red', alpha=.6, size=2) +
    geom_nodelab(aes(x=branch, label=round(posterior, 2)),
                 vjust=-.5, size=3) +
    scale_color_continuous(low='darkgreen', high='red') +
    theme(legend.position=c(.1, .8))
```

在图 5.8 中，我们对树进行了可视化，并将后验概率注释于树对应分枝的中央，也通过最大后验密度（95% HPD）区间呈现出枝长的不确定性。

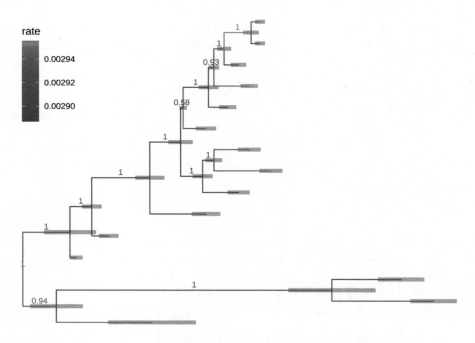

图 5.8　使用 length_0.95_HPD 及后验概率对 BEAST 树进行注释

此示例以红色水平条带来呈现枝长的置信区间（95% HPD），并将后验概率显示于分枝中央。

通过 HyPhy 推断的祖先序列也可以使用 treeio 包进行解析，而每个分支的替换信息则会被自动计算，并存储在系统发育树对象中（即 S4 类对象）。ggtree 包可以利用对象中的这些信息来对树进行注释，如图 5.9 所示。

```
nwk <- system.file("extdata/HYPHY", "labelledtree.tree",
                   package="treeio")
ancseq <- system.file("extdata/HYPHY", "ancseq.nex",
                      package="treeio")
tipfas <- system.file("extdata", "pa.fas", package="treeio")
hy <- read.hyphy(nwk, ancseq, tipfas)ggtree(hy) +
  geom_text(aes(x=branch, label=AA_subs), size=2,
            vjust=-.3, color="firebrick")
```

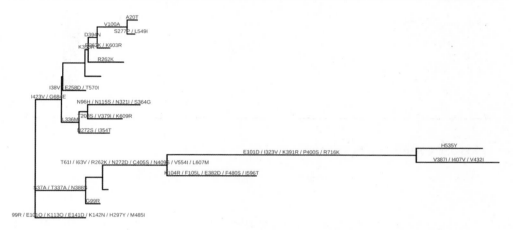

图 5.9　使用根据 HyPhy 推断的祖先序列得出的氨基酸替换信息来对树进行注释

本示例中氨基酸的替代信息被展示在分枝中央。

　　PAML 的 BASEML 和 CODEML 也可用于推断祖先序列，CODEML 还可用于推断选择压力。在使用 treeio 包解析这些信息后，ggtree 包可以将它们整合到同一个树结构中，并对树进行注释，如图 5.10 所示。

```
rstfile <- system.file("extdata/PAML_Codeml", "rst",
                       package="treeio")
mlcfile <- system.file("extdata/PAML_Codeml", "mlc",
                       package="treeio")
ml <- read.codeml(rstfile, mlcfile)
ggtree(ml, aes(color=dN_vs_dS), branch.length='dN_vs_dS') +
  scale_color_continuous(name='dN/dS', limits=c(0, 1.5),
                         oob=scales::squish,
                         low='darkgreen', high='red') +
  geom_text(aes(x=branch, label=AA_subs),
            vjust=-.5, color='steelblue', size=2) +
  theme_tree2(legend.position=c(.9, .3))
```

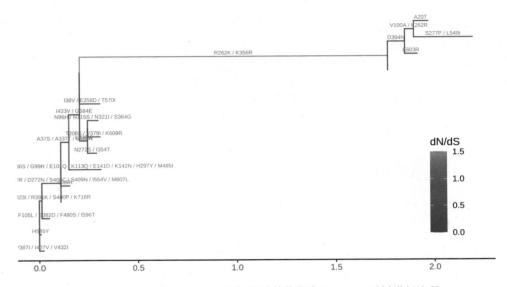

图 5.10　使用 CODEML 推断的氨基酸替换信息及 dN/dS 对树进行注释

此示例中的分支依据 dN/dS 值进行了重新缩放及着色，氨基酸替换信息也被呈现于分支的中央。

ggtree 包解析到的树数据不仅可以用于对系统发育树进行可视化及注释，也可以支持 R 社区中其他包所定义的树及树形对象。ggtree 包在 R 的生态系统中发挥着独特的作用，促进着系统发育学分析，也能被人们轻松地整合到现有的管道及包中。除了可以直接支持树对象，ggtree 包也允许用户绘制含有不同种类的外部数据的树（参见第 7 章，以及本章参考文献 [4]）。

5.4　总结

ggtree 包实现了对系统发育树进行注释的图形语法。用户可以借助 ggplot2 语法，通过组合不同的注释图层来生成复杂的进化树注释。如果对 ggplot2 的使用比较熟悉，那就可以使用 ggtree 包直观且灵活地对进化树进行高度自定义的注释。ggtree 包能收集存储于 treedata 对象中的信息，或者将外部数据关联到树的结构。这就使得我们可以利用系统发育树来进行数据整合分析及比较性研究，极大地拓展系统发育树在其他领域的应用。

5.5 本章练习题

1. 使用图形语法实现进化树可视化的好处是什么呢？

2. 随机生成一棵拥有 20 个叶节点的树。以内向环形布局对齐进行可视化，并以钢青色突出显示 23 号节点的进化枝，使用不同的颜色分别连接 7 号与 11 号、2 号与 15 号、1 号与 9 号叶节点，将它们的粗细比设置为 3:2:1。

3. 使用哪些图层函数可以将进化软件的输出结果注释在进化树上？

参考文献

[1] Yu G, Smith D K, Zhu H, et al. ggtree: an R package for visualization and annotation of phylogenetic trees with their covariates and other associated data[J]. Methods in Ecology and Evolution, 2016, 8(1): 28-36.

[2] Xu S, Dai Z, Guo P, et al. ggtreeExtra: Compact visualization of richly annotated phylogenetic data[J]. Mol Biol Evol, 2021, 38(9): 4039-4042.

[3] Wang L, Lam T T, Xu S, et al. treeio: an R package for phylogenetic tree input and output with richly annotated and associated data[J]. Molecular Biology and Evolution, 2020, 37(2): 599-603.

[4] Yu G, Lam T T, Zhu H, et al. Two methods for mapping and visualizing associated data on phylogeny using ggtree[J]. Mol Biol Evol, 2018, 35(12): 3041-3043.

第 6 章　系统发育树的可视化探索

ggtree[1] 包支持多种在可视化层面对进化树进行操作的方法，包括查看选定进化枝（见图 6.1）、对分类单元进行分组（见图 6.5）、旋转进化枝或进化树（见图 6.6B 和图 6.8）、折叠或展开进化枝（见图 6.2 和图 6.3A）等。表 6.1 所示为进化树操作函数及其说明。

表 6.1　进化树操作函数及其说明

函数名称	说明
collapse()	折叠选中的进化枝
expand()	展开被折叠的进化枝
flip()	交换两个具有相同父节点的进化枝的位置
groupClade()	对进化枝进行分组
groupOTU()	通过追溯回最近共同祖先来对 OTUs 进行分组
identify()	对树进行交互式操作
rotate()	将指定进化枝旋转 180 度
rotate_tree()	将环形布局的树以特定角度进行旋转
scaleClade()	对选定进化枝进行缩放
open_tree()	将树转换为拥有特定开放角度的扇形布局

6.1　查看选定的进化枝

进化枝是指包含单一祖先及其所有后代的单系群。如图 6.1B 所示，我们可以通过 viewClade() 函数选定一个进化枝并对其进行可视化。还有一种方法是将这个进化枝提取出来作为一个新的树对象。这些功能是为了帮助用户探索大型进化树。

```
library(ggtree)
nwk <- system.file("extdata", "sample.nwk", package="treeio")
tree <- read.tree(nwk)
p <- ggtree(tree) + geom_tiplab()
viewClade(p, MRCA(p, "I", "L"))
```

像 viewClade() 这样对进化枝进行操作的函数需要接收一个内部节点编号作为参数。在无法确定内部节点编号时，用户可以使用 MRCA() 函数，如图 6.1 所示，通过提供两个分类单元名称来获取它们的最近共同祖先（MRCA）的节点编号。MRCA() 函数既可以对树使用，又可以对树图使用（即 ggtree 包的输出结果）。tidytree 包中同样提供了一个 MRCA() 函数，用于提取最近共同祖先节点的信息。

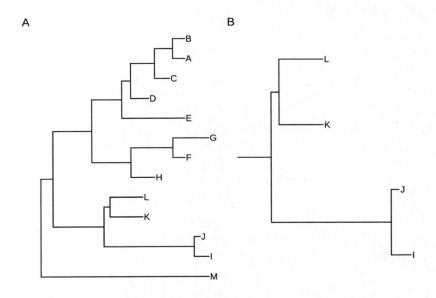

图 6.1　查看选定的进化枝

用于演示 ggtree 是如何支持在可视化层面对进化树进行操作与探究的示例进化树（A）。ggtree 支持对选定的进化枝进行可视化（B）。用户可以通过指定节点编号，或者通过选定两个叶节点，由它们的最近共同祖先来选定进化枝。

6.2 缩小选定的进化枝

ggtree 包还支持通过 scaleClade() 函数来缩小（或者说压放）选定的进化枝。这样，我们就可以保留被压放进化枝的拓扑结构及枝长信息，而节省的空间又可用来重点研究本次感兴趣的进化枝。

```
tree2 <- groupClade(tree, c(17, 21))
p <- ggtree(tree2, aes(color=group)) + theme(legend.position='none') +
  scale_color_manual(values=c("black", "firebrick", "steelblue"))
scaleClade(p, node=17, scale=.1)
```

如果用户想要强调一些重要的进化枝，则可以使用 scaleClade() 函数，将大于 1 的数值传递给 scale 参数，选定的进化枝就会被放大。用户还可以通过 groupClade() 函数为多个选定的进化枝分配不同的进化枝 ID，再根据 ID 为不同的进化枝添加不同的颜色，如图 6.2 所示。

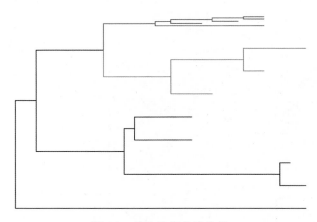

图 6.2 缩小选定的进化枝

进化枝可以被放大（scale > 1）以进行突出显示，或者被缩小以节省空间。

6.3 折叠及展开进化枝

为了强调进化树的某些方面，我们经常会对进化树进行修剪或折叠。ggtree 包支持使用 collapse() 函数折叠选定的进化枝，如图 6.3A 所示。

```
p2 <- p %>% collapse(node=21) +
  geom_point2(aes(subset=(node==21)), shape=21, size=5, fill='green')
p2 <- collapse(p2, node=23) +
  geom_point2(aes(subset=(node==23)), shape=23, size=5, fill='red')
print(p2)
expand(p2, node=23) %>% expand(node=21)
```

此处折叠了两个进化枝，并使用绿色圆圈及红色方块作为符号点进行标记。像这样将过大而无法完整显示出来或不是本次主要研究对象的进化枝折叠起来是一种非常常见的策略。ggtree 包还支持使用 expand() 函数将被折叠的进化枝展开（即取消折叠），来显示物种关系的细节，如图 6.3B 所示。

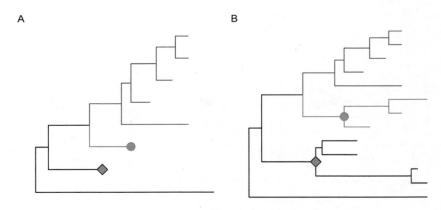

图 6.3　折叠选定的进化枝或展开被折叠的进化枝

我们可以折叠选定的进化枝（A）或重新展开被折叠的进化枝（B）。这是因为 ggtree 包保存着所有的物种关系信息。本示例使用了绿色及红色的符号来代表被折叠的进化枝。

一般我们会使用三角形来指代被折叠的进化枝，ggtree 包也支持这个操作。collapse() 函数提供了一个"mode"参数，默认值为"none"，表示此时进化枝会被折叠为一个叶节点；将"mode"参数设置为"max"时，表示取该进化枝最远端的叶节点的位置构建三角形，如图 6.4A 所示；将"mode"参数设置为"min"时，表示取该进化枝最近端的叶节点的位置构建三角形，如图 6.4B 所示；将"mode"参数设置为"mixed"时，表示综合采用两者的位置构建三角形，如图 6.4C 所示。

```
p2 <- p + geom_tiplab()
node <- 21
collapse(p2, node, 'max') %>% expand(node)
collapse(p2, node, 'min') %>% expand(node)
```

```
collapse(p2, node, 'mixed') %>% expand(node)
```

我们还可以通过传递额外的参数来设置三角形的颜色及透明度，如图 6.4D 所示。

```
collapse(p, 21, 'mixed', fill='steelblue', alpha=.4) %>%
  collapse(23, 'mixed', fill='firebrick', color='blue')
```

我们还可以将 scaleClade() 函数和 collapse() 函数结合起来，对三角形进行缩小操作，如图 6.4E 所示。

```
scaleClade(p, 23, .2) %>% collapse(23, 'min', fill="darkgreen")
```

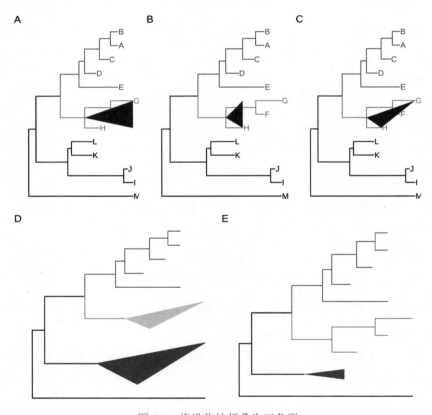

图 6.4　将进化枝折叠为三角形

"max"用于取进化枝最远端的叶节点的位置构建三角形（A）。"min"用于取进化枝最近端的叶节点的位置构建三角形（B）。"mixed"用于综合采用两者的位置构建三角形（C），这样会使折叠后还能保留进化枝的大致形状。设置三角形的线条颜色、填充颜色及透明度（D）。scaleClade() 函数与 collapse() 函数相结合用于缩小三角形以节省空间（E）。

6.4 对分类单元进行分组

groupClade() 函数可以将不同进化枝下的分支和节点分配到不同的组中。该函数通过接收单个内部节点或由多个内部节点组成的向量来对单个或多个进化枝进行聚类。

与之类似，groupOTU() 函数会根据用户指定的操作分类单元（OTUs）分组来将分支与节点分配到不同的组中。这些 OTUs 不一定要在同一个进化枝内，它们可以为单系（在同一个进化枝内）、多系或旁系。groupOTU() 函数在接收由 OTUs 组成的向量或列表后，会追溯至这些 OTUs 的最近共同祖先（MRCA），并将它们聚类在一起，如图 6.5 所示。

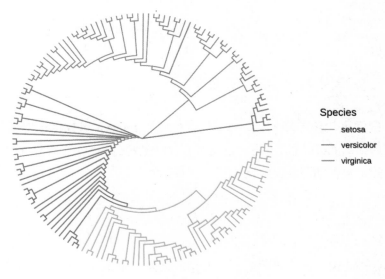

图 6.5　对 OTUs 进行分组

基于 OTUs 之间的关系对其进行分类。用户可以通过指定 OTUs，将它们及其 MRCA 的祖先聚类在一起。

我们可以通过将不同的线条类型、颜色、大小、形状等属性映射到不同分组的分支或节点来对进化树进行注释。

```
data(iris)
rn <- paste0(iris[,5], "_", 1:150)
```

```
rownames(iris) <- rn
d_iris <- dist(iris[,-5], method="man")

tree_iris <- ape::bionj(d_iris)
grp <- list(setosa     = rn[1:50],
            versicolor = rn[51:100],
            virginica  = rn[101:150])

p_iris <- ggtree(tree_iris, layout = 'circular', branch.length='none')
groupOTU(p_iris, grp, 'Species') + aes(color=Species) +
  theme(legend.position="right")
```

我们也可以在树对象的多个水平中完成对分类单元的分组。下面的代码会生成与图 6.5 相同的图像。

```
tree_iris <- groupOTU(tree_iris, grp, "Species")
ggtree(tree_iris, aes(color=Species), layout = 'circular',
        branch.length = 'none') +
  theme(legend.position="right")
```

6.5 对系统发育树结构的探索

为了探索进化树的结构，ggtree 包支持使用 rotate() 函数将选定的进化枝旋转 180 度，如图 6.6B 所示，还可以通过 flip() 函数交换一个内部节点的两个直系子进化枝的位置，如图 6.6C 所示。

```
p1 <- p + geom_point2(aes(subset=node==16), color='darkgreen', size=5)
p2 <- rotate(p1, 16)
flip(p2, 17, 21)
```

大部分进化树的操作函数都是在操作进化枝。而 ggtree 包也提供了一些可以操作整棵树的函数，其中包括可以将矩形布局或环形布局的树转换为扇形布局的 open_tree() 函数，以及将环形布局或扇形布局的树旋转一定角度的 rotate_tree() 函数，如图 6.7 和图 6.8 所示。

我们可以通过 open_tree() 函数将树转换为扇形布局。

```
p3 <- open_tree(p, 180) + geom_tiplab()
print(p3)
```

第 6 章　系统发育树的可视化探索

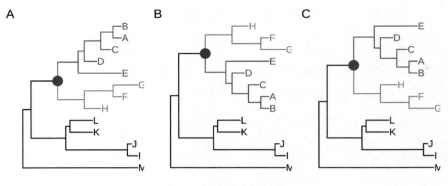

图 6.6　探索进化树结构

我们可以将进化树中的进化枝（此处使用深绿色圆点来表示）（A）旋转180°（B），并且可以交换与其相邻的子进化枝的位置（C）。

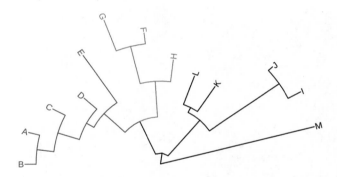

图 6.7　将树转换为扇形布局

我们可以通过 open_tree() 函数以特定开放角度将树转换为扇形布局。

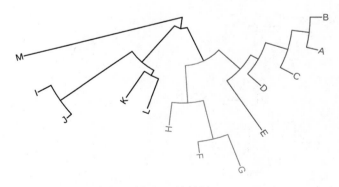

图 6.8　旋转树

环形布局树或扇形布局树可以以任意角度旋转。

我们可以通过 rotate_tree() 函数对树进行旋转。

```
rotate_tree(p3, 180)
```

在下面示例中，我们旋转了 4 个选定的进化枝，如图 6.9 所示。可以看到，我们能轻松地遍历所有内部节点并对其进行逐一旋转。

```
set.seed(2016-05-29)
x <- rtree(50)
p <- ggtree(x) + geom_tiplab()

## nn <- unique(reorder(x, 'postorder')$edge[,1])
## 遍历所有内部节点

nn <- sample(unique(reorder(x, 'postorder')$edge[,1]), 4)

pp <- lapply(nn, function(n) {
    p <- rotate(p, n)
p + geom_point2(aes(subset=(node == n)), color='red', size=3)
})

plot_list(gglist=pp, tag_levels='A', nrow=1)
```

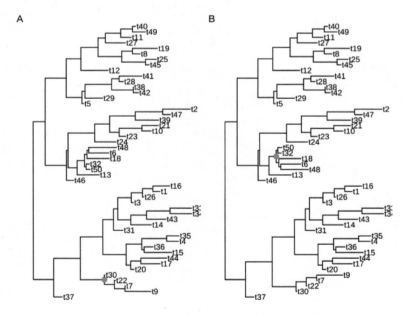

图 6.9　旋转选定的进化枝

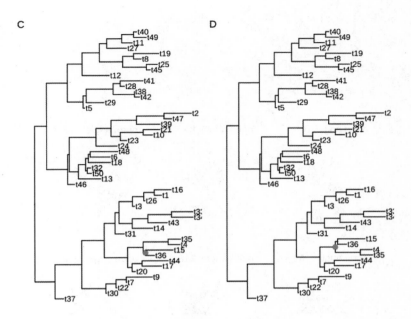

图 6.9 旋转选定的进化枝（续）

随机选择 4 个进化枝进行旋转（用红色圆点标记）。

如图 6.10 所示，我们通过设置不同的开放角度演示了 open_tree() 函数的用法。

```
set.seed(123)
tr <- rtree(50)
p <- ggtree(tr, layout='circular')
angles <- seq(0, 270, length.out=6)

pp <- lapply(angles, function(angle) {
  open_tree(p, angle=angle) + ggtitle(paste("open angle:", angle))
})

plot_list(gglist=pp, tag_levels='A', nrow=2)


set.seed(123)
tr <- rtree(50)
p <- ggtree(tr, layout='circular')
angles <- seq(0, 270, length.out=6)

pp <- lapply(angles, function(angle) {
```

```
    open_tree(p, angle=angle) + ggtitle(paste("open angle:", angle))
})

g <- plot_list(gglist=pp, tag_levels='A', nrow=2)
ggplotify::as.ggplot(g, vjust=-.1,scale=1.1)
```

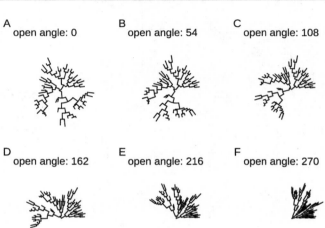

图 6.10　以不同的开放角度将树转换为扇形布局

图 6.11 演示了以不同的角度旋转树的效果。

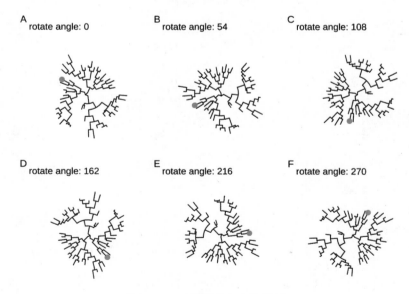

图 6.11　以不同的角度旋转树

通过 identify() 函数，我们也可以实现对树进行交互式的操作。

6.6 总结

一个好的可视化工具除了能帮助用户呈现数据，还能帮助用户探索数据。数据探索可以让用户更好地理解数据，也有助于发现隐藏在数据之下的模式。ggtree 包提供了一系列函数，能够很好地帮助用户在可视化层面操作进化树，同时探索含有关联数据的进化树结构。在进化学背景下探索数据可能有助于用户发现新的系统性模式及提出新的假说。

6.7 本章练习题

1. 随机生成一棵树并将其可视化。任意选择它的一个内部节点，以不同的颜色显示该节点的两个子进化枝。先将任意一个子进化枝放大 3 倍，再将另一个子进化枝进行折叠，折叠后以一个大小为 3 的红色圆形来指代被折叠的进化枝。

2. 随机生成一棵树并将其可视化。任意选择它的一个内部节点，首先将其旋转 180 度，然后将此树以 90 度的开放角度转换为扇形布局，最后将此扇形布局树旋转 270 度。

参考文献

[1] Yu G, Smith D K, Zhu H, et al. ggtree: an R package for visualization and annotation of phylogenetic trees with their covariates and other associated data[J]. Methods in Ecology and Evolution, 2016, 8(1): 28-36.

第 7 章　绘制含有数据的树

我们可以在多个水平将用户数据整合以便对系统发育树进行注释。treeio 包[1]提供了 full_join() 函数，用于将树相关数据整合到系统发育树对象中。tidytree 包支持使用 tidyverse 函数来将树数据关联到进化树。ggtree 包[2]支持即时地将外部数据映射到进化树结构以用于可视化及注释。这几个包关联外部数据的功能看似重复，但其实各有侧重的应用范围。例如，ggtree 包除了支持 treedata 对象，还支持 phylo4d、phyloseq 和 obkData 等多种被设计用于包含特定领域数据的对象。这些对象在被设计时是没有考虑到要支持关联外部数据到对象上的（不能在树对象水平实现关联外部数据）。但我们可以使用 ggtree 包先将这些对象中的树可视化，再在可视化水平关联外部数据[2]。

ggtree 包提供了两种将关联外部数据映射到进化树并进行可视化的通用方法。第一种方法是将外部数据映射到树结构，并在进行树与数据的可视化时将其作为视觉特征呈现出来；第二种方法是根据树结构对数据重新排序，并使用不同的几何对象函数将数据与树并排绘制。这两种方法都能帮助用户将外部数据与系统发育树整合起来，以便在进化生物学背景下进行进一步的探索和比较。我们还开发了 ggtreeExtra 包[3]，其针对矩形布局、环形布局的进化树提供了第二种方法的更好实现。

7.1　将外部数据映射到树结构

ggtree 包提供了一个新的操作符 "%<+%"，可以将注释数据附加到 ggtree 图形对象。只要数据含有内部节点编号的 "node" 列，或者在第一列中提供分类单元标签，就都可以使用 "%<+%" 操作符来将它与 ggtree 对象整合。多个数据集可以被逐步整合。数据被附加到 ggtree 对象后，其中储存的所有信息都会被转换为对应

节点的数值型或分类型属性，并可以在树进行可视化时直接映射到各种图形属性上，如颜色或线条的粗细等。除了直接使用这些属性本身的值来对树进行标注，ggtree 包也支持将它们解析为数学运算式、表情符号或轮廓图等。下面的示例使用了 "%<+%" 操作符对分类单元信息（本示例中的 df_tip_data）与内部节点信息（本示例中的 df_inode_data）进行整合，并将这些数据映射到标签和符号点的颜色及形状上，如图 7.1 所示。ggtree 包也支持解析叶节点数据中的 imageURL 连接，通过链接获取对应物种的在线图片，并将图片作为为叶节点的标签。

```
library(ggimage)
library(ggtree)
library(TDbook)

# 从 TDbook 包中加载 tree_boots、df_tip_data 及 df_inode_data
p <- ggtree(tree_boots) %<+% df_tip_data + xlim(-.1, 4)
p2 <- p + geom_tiplab(offset = .6, hjust = .5) +
    geom_tippoint(aes(shape = trophic_habit, color = trophic_habit,
                size = mass_in_kg)) +
    theme(legend.position = "right") +
    scale_size_continuous(range = c(3, 10))

p2 %<+% df_inode_data +
    geom_label(aes(label = vernacularName.y, fill = posterior)) +
    scale_fill_gradientn(colors = RColorBrewer::brewer.pal(3, "YlGnBu"))
```

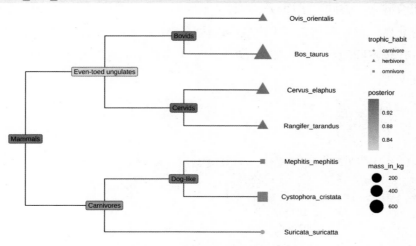

图 7.1 将多个数据集附加到同一棵进化树上

叶节点关联数据（如饮食习惯、体重）及内部节点关联数据（如进化枝后验概率）等外部数据通过 "%<+%" 操作符连接 ggtree 图形对象，并使用它们来对树进行注释。

虽然在 ggtree 包中由 "%<+%" 操作符整合的数据可用于树的可视化,但这些附加于 ggtree 图形对象的数据也可以被转换为包含树及附加数据的 treedata 对象(参见 7.5 小节)。

7.2 基于树的结构将图与树对齐

ggtree 包中 geom_facet() 图层函数及 facet_plot() 函数用于将进化树与用户数据绘制的不同种类的图相关联起来。只需要输入一个数据框并指定所需要绘制的 geom 图层,ggtree 便会在一个新的面板(panel)中绘制对应的图。geom_facet() 图层函数或 facet_plot() 函数是将不同图层与进化树相连的通用解决方案。它们会在内部根据进化树的结构将输入的数据重新排列,并在特定面板上使用对应的几何图层绘制数据。用户可以随意地在多个面板上绘制不同种类的数据,以及在同一个面板上使用不同的几何图层来绘制相同或不同的数据集。

geom_facet() 图层函数支持大多数在 ggplot2 及基于 ggplot2 开发的包中定义的 geom 图层。随着 ggplot2 社区的不断壮大,将来在 ggplot2 包及其拓展包中会有更多的 geom 图层得到实现,这也使得 geom_facet() 图层函数和 facet_plot() 函数呈现数据的功能变得越来越强大。需要注意的是,我们也可以将不同的 geom 图层组合起来,在同一个面板上一起呈示数据,这样就可以将一些更为复杂的数据与进化树一起呈现出来。用户也可以通过逐步添加多个面板以在进化学背景下呈现并比较不同的数据集,如图 7.2 所示。更多的细节可以参考本章参考文献 [2] 的补充文件。

```
library(ggtree)
library(TDbook)

## 从 TDbook 包中加载 tree_nwk、df_info、df_alleles 及 df_bar_data
tree <- tree_nwk
snps <- df_alleles
snps_strainCols <- snps[1,]
snps<-snps[-1,] # 删除株名
colnames(snps) <- snps_strainCols

gapChar <- "?"
snp <- t(snps)
```

```r
lsnp <- apply(snp, 1, function(x) {
        x != snp[1,] & x != gapChar & snp[1,] != gapChar
})
lsnp <- as.data.frame(lsnp)
lsnp$pos <- as.numeric(rownames(lsnp))
lsnp <- tidyr::gather(lsnp, name, value, -pos)
snp_data <- lsnp[lsnp$value, c("name", "pos")]

## 对树进行可视化
p <- ggtree(tree)

## 将样本信息与树相连接,同时添加根据"location"着色的符号
p <- p %<+% df_info + geom_tippoint(aes(color=location))

## 分别使用点图及线图对 SNP 数据和性状数据进行可视化,并将可视化结果与树结构对齐
p + geom_facet(panel = "SNP", data = snp_data, geom = geom_point,
               mapping=aes(x = pos, color = location), shape = '|') +
    geom_facet(panel = "Trait", data = df_bar_data, geom = geom_col,
               aes(x = dummy_bar_value, color = location,
               fill = location), orientation = 'y', width = .6) +
    theme_tree2(legend.position=c(.05, .85))
```

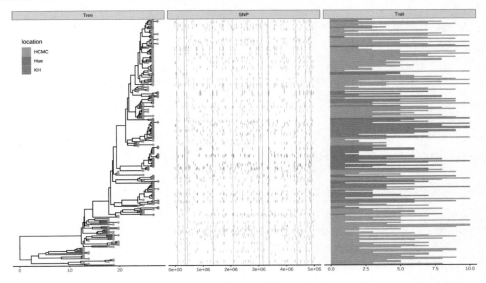

图 7.2 绘制 SNP 及性状数据的示例

"位点(location)"信息在被关联到树后,作为叶节点标签被绘制在进化树上(Tree 面板),SNP 与性状数据分别以点图(SNP 面板)和柱状图(Trait 面板)的形式进行可视化。

geom_facet() 图层函数和 facet_plot() 函数也支持使用 aplot 或 patchwork 包来绘制复合图（参见 7.5 小节）。

由于 geom_facet() 图层函数或 facet_plot() 函数是通过内部调用 ggplot2::facet_grid() 函数实现的，因此只能适用于笛卡尔坐标系。为了在极坐标系下将图与树对齐（如环形布局或扇形布局的树），我们开发了另一款 Bioconductor 包 ggtreeExtra。ggtreeExtra 包提供了 geom_fruit() 图层函数，其功能与 geom_facet() 图层函数的功能类似（详情参见第 10 章）。geom_fruit() 是本章参考文献 [2] 中所提出的第二种方法的更好实现。

7.3 对含有关联矩阵的树进行可视化

ggtree 包提供了 gheatmap() 函数来对含有关联矩阵热图（数值型或分类型）的进化树进行可视化。geom_facet() 图层函数是将数据与树一起绘制的通用解决方案，其同样适用于热图。而 gheatmap() 函数则是专门为同时绘制热图与树设计的，提供了处理列标签及调色盘的快捷方式。除此之外，geom_facet() 图层函数也只支持矩形布局及倾斜布局，而 gheatmap() 函数除了支持这两种布局，还支持环形布局。

下面示例对 H3 流感病毒的进化树及其相关的基因型进行了可视化，如图 7.3A 所示。

```
beast_file <- system.file("examples/MCC_FluA_H3.tree", package="ggtree")
beast_tree <- read.beast(beast_file)

genotype_file <- system.file("examples/Genotype.txt", package="ggtree")
genotype <- read.table(genotype_file, sep="\t", stringsAsFactor=F)
colnames(genotype) <- sub("\\.$", "", colnames(genotype))
p <- ggtree(beast_tree, mrsd="2013-01-01") +
    geom_treescale(x=2008, y=1, offset=2) +
geom_tiplab(size=2)
gheatmap(p, genotype, offset=5, width=0.5, font.size=3,
        colnames_angle=-45, hjust=0) +
    scale_fill_manual(breaks=c("HuH3N2", "pdm", "trig"),
```

```
        values=c("steelblue", "firebrick", "darkgreen"),
        name="genotype")
```

本示例使用 width 参数来调节热图的宽度,以及使用 offset 参数来调节热图与树之间的距离,为叶节点标签节省空间。

对于像本示例一样的时间尺度树,一般我们会使用 theme_tree2() 函数添加 x 轴,但在这里,由于热图只是作为另一个图层存在,因此会改变 x 轴。为了解决这个问题,我们使用 scale_x_ggtree() 函数将 x 轴设置得更为合理,如图 7.3B 所示。

```
p <- ggtree(beast_tree, mrsd="2013-01-01") +
    geom_tiplab(size=2, align=TRUE, linesize=.5) +
    theme_tree2()
gheatmap(p, genotype, offset=8, width=0.6,
        colnames=FALSE, legend_title="genotype") +
    scale_x_ggtree() +
    scale_y_continuous(expand=c(0, 0.3))
```

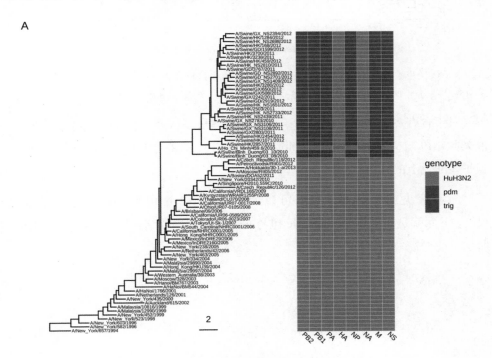

图 7.3 使用 gheatmap() 函数绘制矩阵的示例

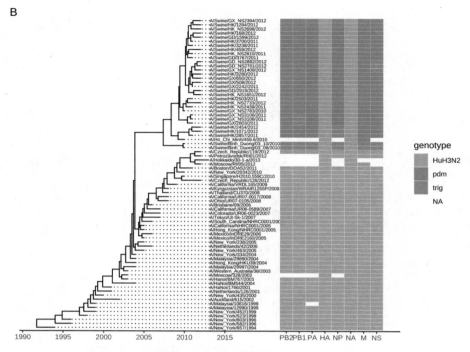

图 7.3 使用 gheatmap() 函数绘制矩阵的示例（续）

本示例使用了 H3 流感病毒树，并对与其关联的基因型表格以热图的形式进行可视化（A）。我们将叶节点对齐，并添加了同时包含分歧时间（树）及基因组片段（热图）的定制 x 轴（B）。

对于含有多个关联矩阵的树，我们可以通过多次调用 gheatmap() 函数来将多个关联矩阵与树对齐。由于 ggplot2 包不允许使用多个 fill 标尺，因此还是会遇到一些配色上的麻烦。

为了解决这个问题，我们可以使用 ggnewscale 包来创建多个新的 fill 标尺[①]。下面就是一个将 ggnewscale 包与 gheatmap() 函数结合使用的示例，如图 7.4 所示。

```
nwk <- system.file("extdata", "sample.nwk", package="treeio")

tree <- read.tree(nwk)
circ <- ggtree(tree, layout = "circular")

df <- data.frame(first=c("a", "b", "a", "c", "d", "d", "a",
                         "b", "e", "e", "f", "c", "f"),
```

---

① 相关讨论请参见"外链资源"文档中第 7 章第 1 条

```
                          second= c("z", "z", "z", "z", "y", "y",
                                   "y", "y", "x", "x", "x", "a", "a"))
rownames(df) <- tree$tip.label

df2 <- as.data.frame(matrix(rnorm(39), ncol=3))
rownames(df2) <- tree$tip.label
colnames(df2) <- LETTERS[1:3]

p1 <- gheatmap(circ, df, offset=.8, width=.2,
               colnames_angle=95, colnames_offset_y = .25) +
    scale_fill_viridis_d(option="D", name="discrete\nvalue")

library(ggnewscale)
p2 <- p1 + new_scale_fill()
gheatmap(p2, df2, offset=15, width=.3,
         colnames_angle=90, colnames_offset_y = .25) +
    scale_fill_viridis_c(option="A", name="continuous\nvalue")
```

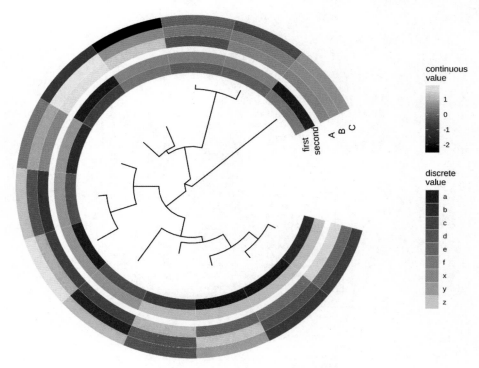

图 7.4　将 ggnewscale 包与 gheatmap() 函数结合使用的示例

多次调用 gheatmap() 函数可以同时展示多个热图与进化树。

7.4 对含有多序列比对结果的树进行可视化

通过输入一棵进化树（ggtree() 的输出）及一个 fasta 文件，我们就可以使用 msaplot() 函数绘制出含有多序列比对结果的树。与 gheatmap() 函数类似，我们也可以通过 width 参数来调节序列比对相对于树的宽度，以及通过 offset 参数来调节热图与树之间的距离，如图 7.5A 所示。

```
library(TDbook)

# 从 TDbook 包中加载 tree_seq_nwk 及 AA_sequence
p <- ggtree(tree_seq_nwk) + geom_tiplab(size=3)
msaplot(p, AA_sequence, offset=3, width=2)
```

我们还可以通过 window 参数来设置仅显示序列比对的特定片段，如图 7.5B 所示。

```
p <- ggtree(tree_seq_nwk, layout='circular') +
    geom_tiplab(offset=4, align=TRUE) + xlim(NA, 12)
msaplot(p, AA_sequence, window=c(120, 200))
```

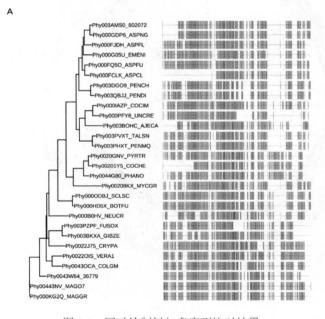

图 7.5　同时绘制树与多序列比对结果

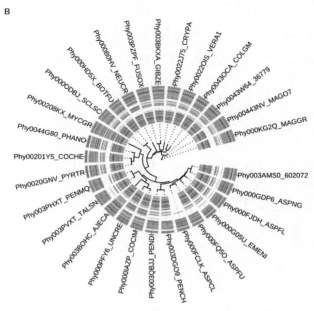

图 7.5　同时绘制树与多序列比对结果（续）

我们将整个 MSA 序列及矩形布局的树一起进行了可视化（A）。附有一段序列窗口的环形布局的树（B）。

为了更好地将多序列比对结果与树和其他关联数据一起可视化，我们开发了 ggmsa 包。它能做到标注序列，并且能使用不同的配色方案为序列着色[4]。ggmsa 包的输出能与 geom_facet() 图层函数和 ggtreeExtra::geom_fruit() 图层函数兼容，因此也可以将它们结合起来，对进化树、多序列比对结果及各种不同类型的关联数据进行可视化，探索这些数据之间的潜在联系。

7.5　复合图

除了使用 geom_facet() 图层函数或 ggtreeExtra::geom_fruit() 图层函数将图自动与树对齐，以及需要使用 gheatmap() 函数和 msaplot() 函数的特殊情况，还可以通过 cowplot 包、patchwork 包、gtable 包[①]或其他的包来构建复合图，但这样需要用户手动将所有的图对齐。这时，ggtree::get_taxa_name() 函数就非常有用了，它可以获取分类单元标签及它们的顺序，帮助用户根据树的结构对数据进行

---
① 相关问题的讨论请参见"外链资源"文档中第 7 章第 2 条

重新排序。为了使拼图的过程不再这么麻烦，我们开发了 aplot 包，它可以自动对 ggplot 对象中的内部数据进行重新排序，并生成与树正确对齐的复合图。

下面示例对一个带有两个关联数据集的树进行了可视化。

```
library(ggplot2)
library(ggtree)

set.seed(2019-10-31)
tr <- rtree(10)

d1 <- data.frame(
    # 不是所有的标签都匹配
    label = c(tr$tip.label[sample(5, 5)], "A"),
    value = sample(1:6, 6))

d2 <- data.frame(
    label = rep(tr$tip.label, 5),
    category = rep(LETTERS[1:5], each=10),
    value = rnorm(50, 0, 3))

g <- ggtree(tr) + geom_tiplab(align=TRUE)

p1 <- ggplot(d1, aes(label, value)) + geom_col(aes(fill=label)) +
    geom_text(aes(label=label, y= value+.1)) +
    coord_flip() + theme_tree2() +
    theme(legend.position='none')

 p2 <- ggplot(d2, aes(x=category, y=label)) +
    geom_tile(aes(fill=value)) + scale_fill_viridis_c() +
    theme_minimal() + xlab(NULL) + ylab(NULL)
```

如果尝试使用 cowplot 包将它们对齐就会发现，复合图并不会像我们预期的那样正确地对齐，如图 7.6A 所示。

```
cowplot::plot_grid(g, p2, p1, ncol=3)
```

而 aplot 包则会帮我们完成所有的操作，自动将所有的子图正确地对齐，如图 7.6B 所示。

```
library(aplot)
p2 %>% insert_left(g) %>% insert_right(p1, width=.5)
```

第 7 章 绘制含有数据的树　145

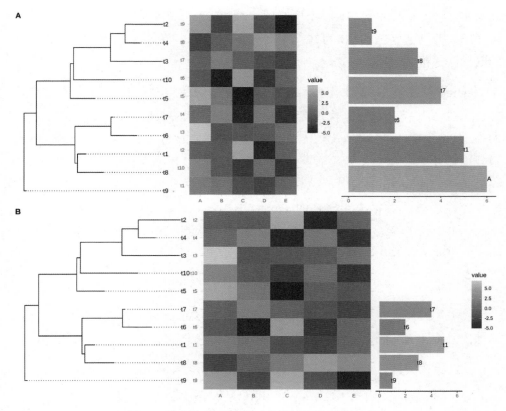

图 7.6　将树与数据并排对齐以创建复合图的示例

cowplot 包只是简单地将子图拼在一起（A），而 aplot 包则会做些额外的工作，以确保这些与树关联的子图能正确地与树对齐。需要注意的是，柱状图中的"A"由于跟树不匹配而没有被绘制出来。

7.6　总结

虽然有很多包都支持对系统发育树进行可视化，但对于那些含有关联数据的树，它们能提供的帮助往往十分有限。也有一些包为存储含有特定领域数据的进化树定义了自己的 S4 类，如 OutbreakTools 包[5] 定义了用于存储带有流行病学数据的树的 obkData，又如 phyloseq 包[6] 定义了用于存储带有微生物组数据的树的 phyloseq。但这些包对存储的外部数据提供的可视化支持并不完善，它们只支持将部分存储于对象中的数据呈现于树上。例如，phyloseq 包并不支持将可视化存储在 phyloseq 对象中的物种丰度信息。而且这些包也并没有提供任何工具来整合外部数据以进行进化树的可视化，更不支持将这些外部数据可视化，并对生成的图根据树的结构进行对齐。

ggtree 包提供了两种用于整合数据的通用解决方法。第一种方法是使用"%<+%"操作符，它可以整合与内外部节点相关联的数据，并将之映射为视觉特征，以便于对树及在 geom_facet() 图层函数或 ggtreeExtra::geom_fruit() 图层函数中使用的数据集进行可视化。第二种方法是使用 geom_facet() 图层函数或 ggtreeExtra::geom_fruit() 图层函数来整合数据，这种方法对输入的数据没有限制，只要存在能够将数据绘制出来的 geom() 图层函数即可（如使用 geom_density_ridges() 图层函数来呈现物种丰度）。用户可以自由组合多个面板，以及在同一个面板中组合不同的几何图层。

我们总结了一些 ggtree 包[2]拥有而许多其他的包没有实现的独特功能。

（1）对与节点或分支关联的数据进行整合，并将之映射为树或其他数据集的视觉特征（见图 7.1）。

（2）支持解析表达式（数学符号或文本格式）、表情符号及图像文件。

（3）geom_facet() 图层函数并没有限制输入数据的类型或绘制数据方式。

（4）支持通过组合不同的 geom() 图层函数来对关联数据进行可视化。

（5）支持在同一个面板上对不同的数据集进行可视化。

（6）支持在 geom_facet() 图层函数中使用"%<+%"操作符整合数据。

（7）支持向特定图层添加更多的注释。

（8）通过对树的可视化、数据整合及图的对齐进行分离，来实现模块化设计。

独特的模块化设计使得 ggtree 从众多包中脱颖而出。我们可以先对存储在树对象中的数据，或者由"%<+%"操作符连接的外部数据与树一起进行可视化，并对树使用多个注释图层实现完全的注释（见图 7.1），再逐步调用 geom_facet() 图层函数，添加多个面板，或者在同一个面板中添加多个图层，使得我们能够绘制出一个完全注释，且带有包含多个图层的复杂数据面板的树。

ggtree 包非常贴合 R 的生态系统，并拓展了现有的系统发育学相关 R 包在对数据与树进行整合及可视化方面的功能。利用 ggtree 包，我们可以使用 phyloseq 对象绘制物种丰度的分布，这在之前是很难做到的。我们还可以通过"%<+%"操作符将外部数据连接到树对象，以及使用 geom_facet() 图层函数将图与树对齐。在现有的流程中加入 ggtree 包，将极大地拓展呈现系统发育相关数据的功能，并拓宽在此方面的应用，特别是对于比较研究来说更是如此。

7.7 本章练习题

随机生成一棵树，并创建一个可以与之合并的外部数据，具体要求如下。

（1）将相关特征映射至树上。

（2）将这棵树与其外部数据生成组合图（geom_facet() 图层函数或 facet_plot() 函数）。

（3）请创建一个能与其合并的热图数据并将其可视化，设置其宽度为树图的一半（width 参数），调整两图的距离。

参考文献

[1] Wang L, Lam T T, Xu S, et al. treeio: an R package for phylogenetic tree input and output with richly annotated and associated data[J]. Molecular Biology and Evolution, 2020, 37(2): 599-603.

[2] Yu G, Lam T T, Zhu H, et al. Two methods for mapping and visualizing associated data on phylogeny using ggtree[J]. Mol Biol Evol, 2018, 35(12): 3041-3043.

[3] Xu S, Dai Z, Guo P, et al. ggtreeExtra: Compact visualization of richly annotated phylogenetic data[J]. Mol Biol Evol, 2021, 38(9): 4039-4042.

[4] Yu G. Using ggtree to visualize data on tree-like structures[J]. Current Protocols in Bioinformatics, 2020, 69(1): e96.

[5] Jombart T, Aanensen D M, Baguelin M, et al. outbreakTools: a new platform for disease outbreak analysis using the R software[J]. Epidemics, 2014, 7: 28-34.

[6] McMurdie P J H S. phyloseq: an R package for reproducible interactive analysis and graphics of microbiome census data[J]. 2013.

第 8 章　使用轮廓图和子图注释进化树

8.1 使用图像注释进化树

通常我们会以文字的形式，也就是通过展示分类单元的名称来添加分类单元标签。如果此处的文字是一个图像的文件名（本地文件或远程文件皆可），ggtree 包就会读取图像并将实际图像展示为分类单元的标签，如图 8.1 所示。有了 ggimage 包提供的支持，geom_tiplab() 图层函数和 geom_nodelab() 图层函数还可用于渲染轮廓图。

iTOL[1] 和 EvolView[2] 等线上工具均支持在进化树上展示子图，但它们仅支持展示柱状图或饼图。有时，用户可能想以一些其他的可视化方法来对节点关联的数据进行可视化，如小提琴图[3]、维恩图[4]、序列标识图等，并将它们展示在进化树上。此时，这些工具就不能很好地提供帮助。而在 ggtree 包中，子图会被先输出为图像文件，再被作为标签添加至对应节点，这样用户便可以在进化树上添加任何种类的子图。

```
library(ggimage)
library(ggtree)

nwk <- paste0("((((bufonidae, dendrobatidae), ceratophryidae),",
        "(centrolenidae, leptodactylidae)), hylidae);")

imgdir <- system.file("extdata/frogs", package = "TDbook")

x = read.tree(text = nwk)
ggtree(x) + xlim(NA, 7) + ylim(NA, 6.2) +
    geom_tiplab(aes(image=paste0(imgdir, '/', label, '.jpg')),
            geom="image", offset=2, align=2, size=.2)   +
```

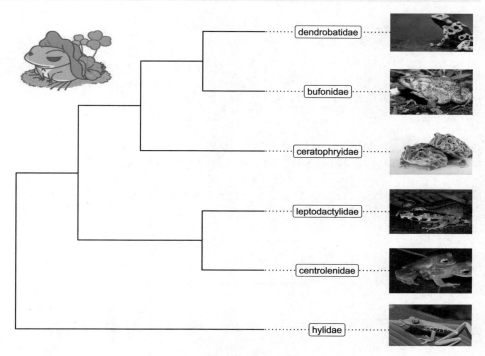

图 8.1　使用图像作为分类单元标签

用户需要指定 geom = "image"，并将图像文件名映射到 image 美学属性。

8.2　使用 phylopic 注释进化树

phylopic 网站中包含 3200 多张轮廓图，几乎涵盖了所有的生命形式。ggtree 包支持使用 phylopic[①] 来对进化树进行注释，只需设置 geom="phylopic"，并将 phylopic UID 映射到 image 美学属性即可。ggimage 包还支持通过学名获取 phylopic UID，这样就可以非常方便地使用 phylopic 来注释进化树。下面的示例通过叶节点标签获取了 phylopic UID，并将对应的 phylopic 作为第二层叶节点标签对树进行标注。最重要的是，我们还可以使用数值或分类变量对图像进行着色或缩放。本示例使用体重来对图像的颜色进行调整，如图 8.2 所示。

---
① 相关推文请参见"外链资源"文档中第 8 章第 1 条

```
library(ggtree)
newick <- paste0("((Pongo_abelii,(Gorilla_gorilla_gorilla,(Pan_paniscus,",
         "Pan_troglodytes)Pan,Homo_sapiens)Homininae)Hominidae,",
         "Nomascus_leucogenys)Hominoidea;")

tree <- read.tree(text=newick)

d <- ggimage::phylopic_uid(tree$tip.label)
d$body_mass <- c(52, 114, 47, 45, 58, 6)

p <- ggtree(tree) %<+% d +
  geom_tiplab(aes(image=uid, colour=body_mass), geom="phylopic",
              offset=2.5) +
  geom_tiplab(aes(label=label), offset = .2) + xlim(NA, 7) +
  scale_color_viridis_c()
```

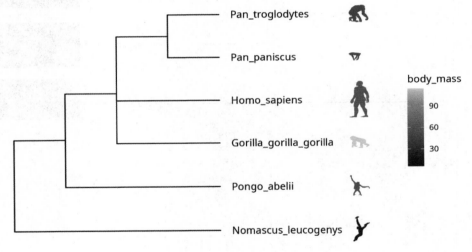

图 8.2 使用 phylopic 作为分类单元标签

ggtree 包会自动通过提供的 UID 下载对应的 phylopic 图像，也支持使用分类变量或数值变量对图像着色。

8.3 使用子图注释进化树

ggtree 包中的 geom_inset() 图层函数可用于将子图添加到系统发育树中。其需要一个由对应节点编号命名的 ggplot 图形对象（任何类型的图表均可以）组成的命名列表作为输入。用户还可以通过 ggplotify 将其他函数（甚至是由 Base 绘图系统实现的函数）生成的图转换为 ggplot 对象，再在 geom_inset() 图层函数中进行使用。为了更加方便地添加柱状图和饼图（如呈现祖先重构结果的汇总数据等），ggtree 包还提供了 nodepie() 函数和 nodebar() 函数来创建由饼图或柱状图构成的列表。

8.3.1 使用柱状图进行注释

本示例使用 ape::ace() 函数估计祖先特征状态。我们将对应的似然值以堆叠柱状图的形式进行可视化，再使用 geom_inset() 图层函数将其覆盖到进化树的内部节点上，如图 8.3A 所示。

```
library(phytools)
data(anoletree)
x <- getStates(anoletree,"tips")
tree <- anoletree

cols <- setNames(palette()[1:length(unique(x))],sort(unique(x)))
fitER <- ape::ace(x,tree,model="ER",type="discrete")
ancstats <- as.data.frame(fitER$lik.anc)
ancstats$node <- 1:tree$Nnode+Ntip(tree)

## cols 参数表明哪些列用于存储 stats 信息
bars <- nodebar(ancstats, cols=1:6)
bars <- lapply(bars, function(g) g+scale_fill_manual(values = cols))

tree2 <- full_join(tree, data.frame(label = names(x), stat = x ),
                   by = 'label')
p <- ggtree(tree2) + geom_tiplab() +
    geom_tippoint(aes(color = stat)) +
    scale_color_manual(values = cols) +
    theme(legend.position = "right") +
    xlim(NA, 8)
p1 <- p + geom_inset(bars, width = .08, height = .05, x = "branch")
```

参数 x 是指放置子图的位置，通过将其设置为 "node" 或 "branch" 可以将子图放置于节点或分支上，还可以通过设置参数 hjust 和 vjust 分别对其进行水平及垂直方向的调整。插入子图的大小则可以通过设置参数 width 和 height 来调整。

8.3.2 使用饼图进行注释

用户也可以先使用 nodepie() 函数生成由饼图组成的列表，再将这些图表放置于对应的节点上进行注释，如图 8.3B 所示。nodebar() 函数和 nodepie() 函数都可以通过设置 alpha 参数来改变图表的透明度。

```
pies <- nodepie(ancstats, cols = 1:6)
pies <- lapply(pies, function(g) g+scale_fill_manual(values = cols))
p2 <- p + geom_inset(pies, width = .1, height = .1)

plot_list(p1, p2, guides='collect', tag_levels='A')
```

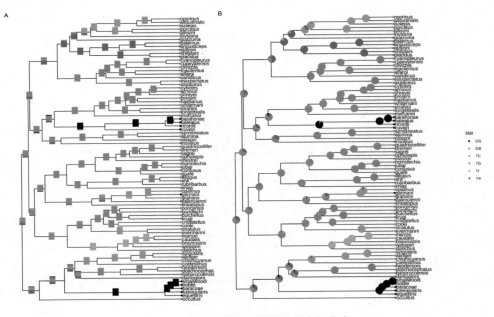

图 8.3 使用柱状图或饼图注释内部节点

使用柱状图（A）或饼图（B）呈现内部节点的汇总数据。

8.3.3 使用多种不同类型的图表进行注释

geom_inset() 图层函数需要接收一个由 ggplot 图形对象组成的列表，而这些输入的对象并不局限于饼图或柱状图，它们可以是任意类型的图表，也可以是这些图表的组合。geom_inset() 图层函数也不单单可以用于呈现祖先数据。我们也

第 8 章　使用轮廓图和子图注释进化树

可以用它来对不同类型的、与树中指定节点关联的数据进行可视化。下面示例通过组合饼图及柱状图来对树进行注释，如图 8.4 所示。

```
pies_and_bars <- pies
i <- sample(length(pies), 20)
pies_and_bars[i] <- bars[i]
p + geom_inset(pies_and_bars, width=.08, height=.05)
```

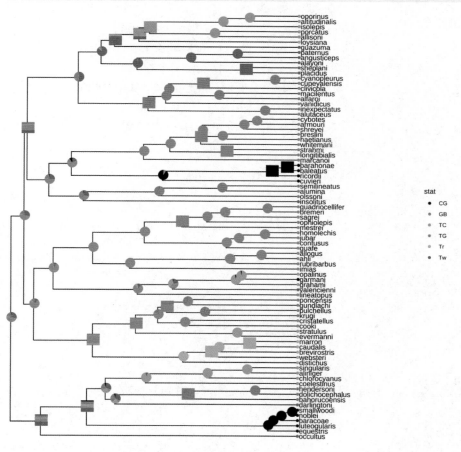

图 8.4　使用不同类型的子图注释内部节点

8.4　玩转 phylomoji

phylomoji 是由表情符号组成的系统发育树，其寓教于乐[①]，能帮助我们更好

---
① 更多 phylomoji 的示例请参见"外链资源"文档中第 8 章第 2 条

地理解进化的概念，非常适合在教学中使用。ggtree 包自 2015 年[①]起便支持生成 phylomoji。本示例通过使用 ggtree 包重新绘制图 8.5 中的 phylomoji[②]，结果如图 8.6 所示。

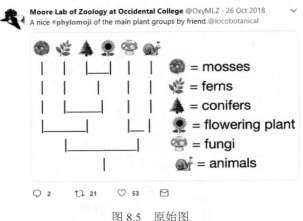

图 8.5　原始图

```
library(ggplot2)
library(ggtree)

tt = '((snail,mushroom),(((sunflower,evergreen_tree),leaves),green_
salad));'
tree = read.tree(text = tt)
d <- data.frame(label = c('snail','mushroom', 'sunflower',
                          'evergreen_tree','leaves', 'green_salad'),
                group = c('animal', 'fungi', 'flowering plant',
                          'conifers', 'ferns', 'mosses'))

p <- ggtree(tree, linetype = "dashed", size=1, color='firebrick') %<+%
d +
    xlim(0, 4.5) + ylim(0.5, 6.5) +
    geom_tiplab(parse="emoji", size=15, vjust=.25) +
    geom_tiplab(aes(label = group), geom="label", x=3.5, hjust=1)
```

需要注意的是，最终呈现的结果可能取决于用户系统中安装的表情符号字体[③]。

[①] 相关细节请参见"外链资源"文档中第 8 章第 3 条
[②] 此推文链接请参见"外链资源"文档中第 8 章第 4 条
[③] 作者在系统中安装的是谷歌 noto 表情符号字体

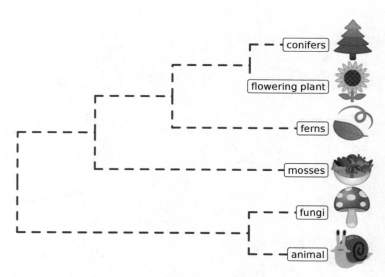

图 8.6 将节点标签解析为表情符号

ggtree 包支持将文本（如节点标签）解析为表情符号。

通过 ggtree 包，我们可以很轻松地生成 phylomoji，其中的表情符号被视作如 "abc" 一般的文本。我们可以使用表情符号对分类单元、进化枝进行标注，也可以随意对表情符号进行旋转或着色。这个功能主要通过 emojifont 包在内部提供支持来实现。

8.4.1 在环形布局或扇形布局的树中使用表情符号

我们也可以在环形布局或扇形布局的树中使用表情符号，如图 8.7 所示。

```
p <- ggtree(tree, layout = "circular", size=1) +
  geom_tiplab(parse="emoji", size=10, vjust=.25)
print(p)

## 扇形布局
p2 <- open_tree(p, angle=200)
 print(p2)

p2 %>% rotate_tree(-90)
```

我们可以在科研文章[5]中找到另一个使用 ggtree 包和 emojifont 包生成的由植物表情符号构成的系统发育树的示例。

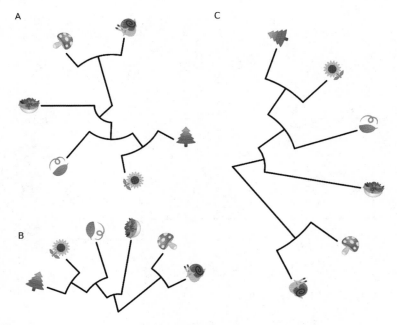

图 8.7 在环形布局或扇形布局的树中使用表情符号

8.4.2 使用表情符号作为进化枝标签

geom_cladelab() 图层函数同样支持将进化枝标签解析为表情符号。例如，在流感病毒的系统发育树中，我们可以使用表示宿主物种的表情符号来标注进化枝，如图 8.8 所示。

```
set.seed(123)
tr <- rtree(30)

dat <- data.frame(
        node = c(41, 53, 48),
        name = c("chicken", "duck", "family")
    )

p <- ggtree(tr) +
    xlim(NA, 5.2) +
    geom_cladelab(
        data = dat,
        mapping = aes(
```

```
            node = node,
            label = name,
            color = name
        ),
        parse = "emoji",
        fontsize = 12,
        align = TRUE,
        show.legend = FALSE
    ) +
    scale_color_manual(
        values = c(
            chicken="firebrick",
            duck="steelblue",
            family = "darkkhaki"
        )
    )
p
```

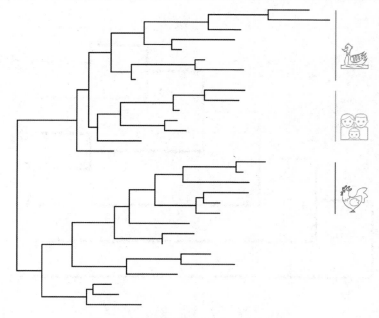

图 8.8　使用表情符号作为进化枝标签

8.4.3　Apple 彩色表情符号

虽然 R 的绘图工具并不支持 macOS 上的 AppleColorEmoji 字体，但方法总比困难多。我们可以先将树图以 svg 的形式保存，再在 Safari 中对其进行渲染。如图 8.9 所示，叶节点标签被 Safari 解析为 AppleColorEmoji 字体的表情符号。

```
library(ggtree)
tree_text <- paste0("(((((cow, (whale, dolphin)), (pig2, boar)),",
                    "camel), fish), seedling);")
x <- read.tree(text=tree_text)
library(ggimage)
p <- ggtree(x, size=2) + geom_tiplab(size=20, parse='emoji') +
    xlim(NA, 7) + ylim(NA, 8.5)

svglite::svglite("emoji.svg", width = 10, height = 7)
print(p)
dev.off()

# 或者使用 grid.export() 函数以达到上述代码的相同效果
# ps = gridSVG::grid.export("emoji.svg", addClass=T)
```

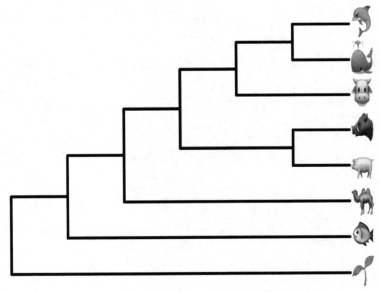

图 8.9　在 ggtree 中使用 Apple 彩色表情符号

8.4.4　使用 ASCII Art 呈现 phylomoji

我们也可以以 ASCII Art 的形式呈现 phylomoji，具体内容请读者参见附录 B 中的 B.5 小节。

8.5 总结

当叶节点标签、内部节点标签及进化枝标签在内的多种标签可以被分别解析为图像文件名、plotmath 表达式或表情符号名时，ggtree 包可以将它们分别解析为图像、数学表达式或表情符号。这不仅十分有趣，在科研中也能为用户提供不少帮助。通过在系统发育树中添加图像，我们可以更好地呈现与物种相关的特性，如形态结构、解剖结构，甚至是大分子结构等。不仅如此，ggtree 包还可以对统计推断（如生物地理范围重建和后验概率分布等）或节点关联数据进行总结归纳，并以子图的形式呈现在进化树上。

8.6 本章练习题

1. 随机生成一棵树，并使用任意本地图片作为叶节点标签来注释它。

2. 随机生成一棵树，使用 ggplot2 与 R 的内置数据集 mtcars 任意生成等于内部节点数量的子图，并将这些子图注释于生成的进化树的内部节点上。

3. 参考外链资源第 8 章第 2 条链接中的内容，创造属于自己的 Phylomoji。

参考文献

[1] Letunic I, Bork P. Interactive Tree Of Life (iTOL): an online tool for phylogenetic tree display and annotation[J]. Bioinformatics, 2007, 23(1): 127-128.

[2] He Z, Zhang H, Gao S, et al. evolview v2: an online visualization and management tool for customized and annotated phylogenetic trees[J]. Nucleic Acids Res, 2016, 44(W1): W236-W241.

[3] Grubaugh N D, Ladner J T, Kraemer M, et al. Genomic epidemiology reveals multiple introductions of Zika virus into the United States[J]. Nature, 2017, 546(7658): 401-405.

[4] Lott S C, Voss B, Hess W R, et al. coVenntree: a new method for the comparative analysis of large datasets[J]. Front Genet, 2015, 6: 43.

[5] Escudero M, Wendel J F. The grand sweep of chromosomal evolution in angiosperms[J]. New Phytol, 2020, 228(3): 805-808.

第 3 篇
ggtree 拓展包

第 9 章 对其他树形对象使用 ggtree 包

9.1 使用 ggtree 包绘制系统发育树对象

treeio 包[1]支持从多个软件输出结果中解析出进化推论数据，并将外部数据关联到树结构上，这使得 treeio 可以作为将进化数据带入 R 社区的基础。ggtree 包[2]能够与 treeio 包无缝协作，通过对树关联数据进行可视化来注释树。ggtree 包可以作为在树可视化及注释方面的通用工具，并且也非常符合 R 包的生态系统。ggtree 包支持大部分由其他 R 包定义的 S3 类或 S4 类树对象，如 phylo、multiPhylo、phylo4、phylo4d、phyloseq 和 obkData。通过 ggtree 包能轻松生成更为复杂的树图，这是其他包很难或根本不能做到的。例如，phyloseq 包无法像 ggtree 包一样实现对 phyloseq 对象更为复杂的可视化（图 9.4）。除此之外，ggtree 包还拓展了将外部数据关联到这些树对象上的可能性[3]。

9.1.1 phylo4 对象和 phylo4d 对象

phylo4 对象及 phylo4d 对象是在 phylobase 包中定义的。其中，phylo4 对象是 phylo 对象的 S4 版本，而 phylo4d 则是在 phylo4 的基础上添加了一个含有性状数据的数据框。phylobase 包提供了一个 plot() 函数，并在内部调用 treePlot() 函数来对含有数据的树进行可视化。但是 plot() 函数也存在着一些限制，它只能将数值型的进化树关联数据绘制为气泡图，也不支持生成图例。phylobase 包并没有实现对分类型的值进行可视化的函数，也不支持将这些关联数据转换为诸如颜色、大小、形状等视觉特征。虽然利用 phylobase 包可以做到使用关联数据为进化树着色，但需要用户手动提取数据，并将它们映射到颜色向量，再将颜色向量传递给 plot() 函数。这是一个非常乏味的过程，也非常容易出错，因为用户需要保证颜色向量的顺序与对象中存储的分支列表的顺序相一致。

ggtree 包支持对 phylo4d 对象的可视化，同时，存储在 phylo4d 对象中的所有关联数据都可以直接用来对树进行注释，如图 9.1 所示。

```r
library(phylobase)
data(geospiza_raw)
g1 <- as(geospiza_raw$tree, "phylo4")
g2 <- phylo4d(g1, geospiza_raw$data, missing.data="warn")

d1 <- data.frame(x = seq(1.1, 2, length.out = 5),
                 lab = names(geospiza_raw$data))

p1 <- ggtree(g2) +
   geom_tippoint(aes(size = wingL),   x = d1$x[1], shape = 1) +
   geom_tippoint(aes(size = tarsusL), x = d1$x[2], shape = 1) +
   geom_tippoint(aes(size = culmenL), x = d1$x[3], shape = 1) +
   geom_tippoint(aes(size = beakD),   x = d1$x[4], shape = 1) +
   geom_tippoint(aes(size = gonysW),  x = d1$x[5], shape = 1) +
   scale_size_continuous(range = c(3,12), name="") +
   geom_text(aes(x = x, y = 0, label = lab), data = d1, angle = 45) +
   geom_tiplab(offset = 1.3) + xlim(0, 3) +
   theme(legend.position = c(.1, .75))

## 用户可以通过 as.treedata(g2) 将 g2 转换为 treedata 对象，
## 并使用 get_tree_data() 函数来提取进化树关联数据

p2 <- gheatmap(ggtree(g1), data=geospiza_raw$data, colnames_angle=45) +
   geom_tiplab(offset=1) + hexpand(.2) +
   theme(legend.position = c(.1, .75))

aplot::plot_list(p1, p2, ncol=2, tag_levels='A')
```

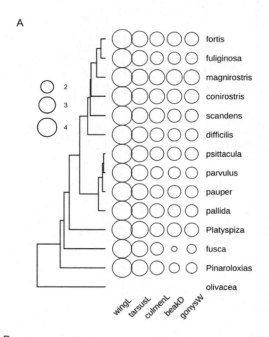

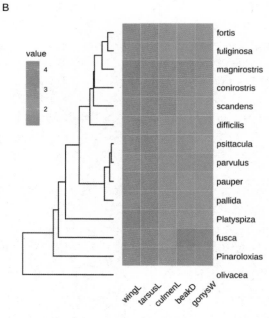

图 9.1 使用 ggtree 包对 phylo4d 对象中的数据进行可视化

重现 phylobase 包中 plot() 函数的输出结果（A）。phylobase 包并不支持以热图的形式对性状数据进行可视化，而 ggtree 包则支持（B）。

9.1.2 phylog 对象

phylog 对象是在 ade4 包中定义的。此包是专门为分析生态学数据设计的，同时提供了 newick2phylog() 函数、hclust2phylog() 函数及 taxo2phylog() 函数，通过 Newick 字符串、层次聚类结果及分类信息来构建系统发育树，并生成 phylog 对象。ggtree 包也支持对 phylog 对象的可视化，如图 9.2 所示。

```
library(ade4)
data(taxo.eg)
tax <- as.taxo(taxo.eg[[1]])
names(tax) <- c("genus", "family", "order")
print(tax)
```

```
##         genre  famille  ordre
## esp3    g1     fam1     ORD1
## esp1    g1     fam1     ORD1
## esp2    g1     fam1     ORD1
## esp4    g1     fam1     ORD1
## esp5    g1     fam1     ORD1
## esp6    g1     fam1     ORD1
## esp7    g1     fam1     ORD1
## esp8    g2     fam2     ORD2
## esp9    g3     fam2     ORD2
## esp10   g4     fam3     ORD2
## esp11   g5     fam3     ORD2
## esp12   g6     fam4     ORD2
## esp13   g7     fam4     ORD2
## esp14   g8     fam5     ORD2
## esp15   g8     fam5     ORD2
```

```
tax.phy <- taxo2phylog(as.taxo(taxo.eg[[1]]))
print(tax.phy)
```

```
## Phylogenetic tree with 15 leaves and 16 nodes
## $class: phylog
## $call: taxo2phylog(taxo = as.taxo(taxo.eg[[1]]))
## $tre: ((((esp3,esp1,esp2,esp4,e...15)l1g8)l2fam5)l3ORD2)Root;
```

```
## 
##         class     length
## $leaves numeric   15
## $nodes  numeric   16
## $parts  list      16
## $paths  list      31
## $droot  numeric   31
##         content
## $leaves length of the first preceeding adjacent edge
## $nodes  length of the first preceeding adjacent edge
## $parts  subsets of descendant nodes
## $paths  path from root to node or leave
## $droot  distance to root
```

```
ggtree(tax.phy) + geom_tiplab() +
  geom_nodelab(geom='label') + hexpand(.05)
```

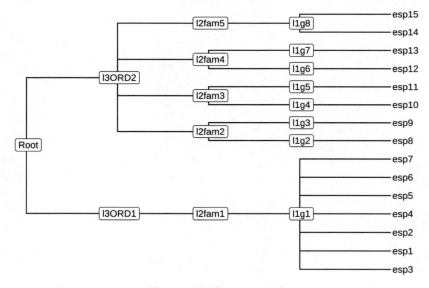

图 9.2 可视化 phylog 对象

9.1.3 phyloseq 对象

phyloseq 包中定义的 phyloseq 类是专门为了存储包括系统发育树、与树关联的样本数据及分类学分配信息（taxonomy assignment）在内的微生物组数据设计的。phyloseq 包可以从许多比较常见的管道的输出中导入数据，如 QIIME [4]、mothur[5] dada2[6] 及 PyroTagger[7] 等。ggtree 包支持对存储在 phyloseq 对象中的进化树进行可视化，其中的关联数据也可以用来对树进行注释，如图 9.3 所示。

```
library(phyloseq)
library(scales)

data(GlobalPatterns)
GP <- prune_taxa(taxa_sums(GlobalPatterns) > 0, GlobalPatterns)
GP.chl <- subset_taxa(GP, Phylum=="Chlamydiae")

ggtree(GP.chl) +
  geom_nodelab(aes(label=label), hjust=-.05, size=3.5) +
  geom_tiplab(aes(label=Genus), hjust=-.3) +
  geom_point(aes(x=x+hjust, color=SampleType, shape=Family,
             size=Abundance), na.rm=TRUE) +
  scale_size_continuous(trans=log_trans(5)) +
  theme(legend.position="right")
```

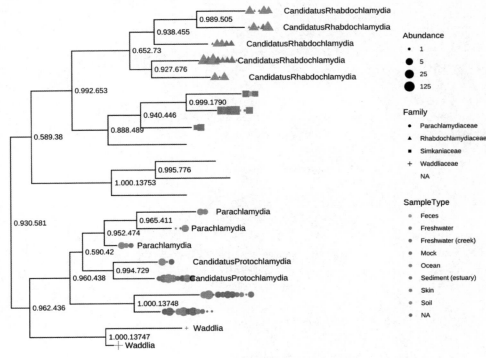

图 9.3　对 phyloseq 包中存储的对象进行可视化

图 9.3 重现了 phyloseq::plot_tree() 函数的输出结果。用户在使用 ggtree 包时会发现，它在对微生物组数据的可视化及进一步注释方面相当适用，这是因为 ggtree 包支持使用图形语法进行高层次的注释，并能添加很多 phyloseq 包不能添加的进化树关联数据图层。

在图 9.4 中，我们使用了 phyloseq 包中提供的微生物组数据，并使用密度脊线图来对物种丰度数据进行可视化。geom_facet() 图层函数会自动根据树的结构重新对丰度数据进行排序，并使用指定的 geom() 图层函数对数据进行可视化（本示例中为 geom_density_ridges() 图层函数），以及将密度曲线与树对齐。值得注意的是，使用 ggtree() 函数可以获取存储在 phyloseq 对象中的数据，并可以在树的可视化时直接使用这些数据（本示例中，我们使用物种所属的门的信息来为叶节点及密度脊线着色）。本示例的源代码率先发表于本章参考文献 [3] 的补充文件中。

```
library(ggridges)

data("GlobalPatterns")
GP <- GlobalPatternsGP <- prune_taxa(taxa_sums(GP) > 600, GP)
sample_data(GP)$human <- get_variable(GP, "SampleType") %in%
  c("Feces", "Skin")

mergedGP <- merge_samples(GP, "SampleType")
mergedGP <- rarefy_even_depth(mergedGP,rngseed=394582)
mergedGP <- tax_glom(mergedGP,"Order")

melt_simple <- psmelt(mergedGP) %>%
  filter(Abundance < 120) %>%
  select(OTU, val=Abundance)

ggtree(mergedGP) +
  geom_tippoint(aes(color=Phylum), size=1.5) +
  geom_facet(mapping = aes(x=val,group=label,
                           fill=Phylum),
             data = melt_simple,
             geom = geom_density_ridges,
             panel="Abundance",
             color='grey80', lwd=.3) +
  guides(color = guide_legend(ncol=1)
```

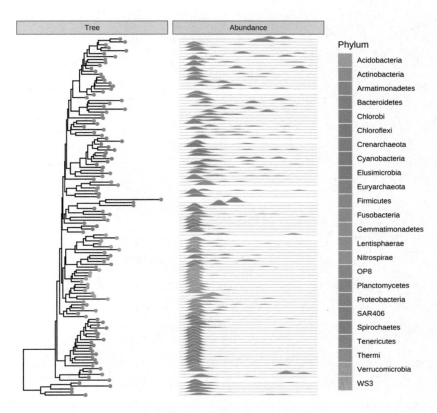

图 9.4　绘制系统发育树及 OTU 丰度密度

根据物种所属门的不同来对叶节点进行着色，同时以密度脊线图的形式对它们在样本间的对应丰度进行可视化，并根据树的结构进行排序。

9.2　使用 ggtree 包绘制树状图

树状图（Dendrogram）常用于展示层次聚类结果及分类树或回归树。在 R 语言中，我们可以使用 hclust() 函数进行层次聚类的计算。

```
hc <- hclust(dist(mtcars))
hc
```

```
## 
## Call:
## hclust(d = dist(mtcars))
```

```
## 
## Cluster method   : complete
## Distance         : euclidean
## Number of objects: 32
```

hclust 对象描述了在聚类过程中产生的树。我们可以将它转换为以深度嵌套的列表来存储树的 dendrogram 对象。

```
den <- as.dendrogram(hc)
den
```

```
## 'dendrogram' with 2 branches and 32 members total, at height
425.3
```

ggtree 包支持 R 社区中定义的大多数层次聚类对象，如 hclust、dendrogram，cluster 包中定义的 agnes、diana、twins，以及 pvclust 包中定义的 pvclust 对象。用户可以通过 ggtree(object) 指令对其树结构进行可视化，并使用其他图层或功能自定义树图或为树图添加注释。

ggtree 包中的 layout_dendrogram() 函数用于将树设置为自上而下的布局，而 theme_dendrogram() 函数用于显示树的高度（类似于系统发育树可视化中 theme_tree2() 函数的功能），如图 9.5 所示，也可参见本章参考文献 [8] 中的示例。

```
clus <- cutree(hc, 4)
g <- split(names(clus), clus)

p <- ggtree(hc, linetype='dashed')
clades <- sapply(g, function(n) MRCA(p, n))

p <- groupClade(p, clades, group_name='subtree') + aes(color=subtree)
d
<- data.frame(label = names(clus),
              cyl = mtcars[names(clus), "cyl"])

p %<+% d +
  layout_dendrogram() +
  geom_tippoint(aes(fill=factor(cyl), x=x+.5),
                size=5, shape=21, color='black') +
  geom_tiplab(aes(label=cyl), size=3, hjust=.5, color='black') +
  geom_tiplab(angle=90, hjust=1, offset=-10, show.legend=FALSE) +
```

```
scale_color_brewer(palette='Set1', breaks=1:4) +
theme_dendrogram(plot.margin=margin(6,6,80,6)) +
theme(legend.position=c(.9, .6))
```

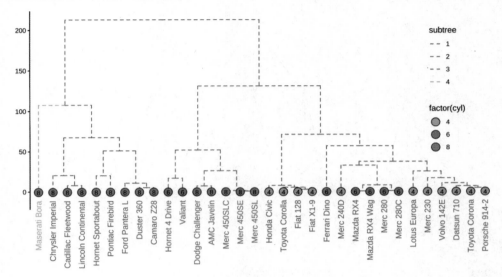

图 9.5 对树状图进行可视化

本示例使用了 cutree() 函数将树分成不同组，并使用 groupClade() 函数依据此信息对树进行分组。树以经典的自上而下的布局进行展示，并依据分组信息对分支进行着色，同时叶节点由汽缸的数量进行着色及标注。

9.3 使用 ggtree 包绘制树形网络图

树形网络图（Tree Graph，如一个 igraph 对象）可以通过 treeio 包中的 as.phylo() 函数转换为 phylo 对象。ggtree 包支持直接对树形网络图表进行可视化，如图 9.6 所示。需要注意的是，目前 ggtree 包并不支持所有 igraph 对象的可视化，它仅支持对树形网络图形式的 igraph 对象的可视化。

```
library(igraph)
g <- graph.tree(40, 3)
arrow_size <- unit(rep(c(0, 3), times = c(27, 13)), "mm")
ggtree(g, layout='slanted', arrow = arrow(length=arrow_size)) +
  geom_point(size=5, color='steelblue', alpha=.6) +
  geom_tiplab(hjust=.5,vjust=2) + layout_dendrogram()
```

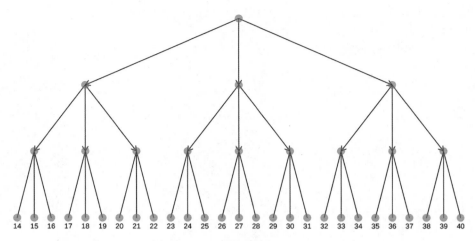

图 9.6　对树形网络图进行可视化

本示例使用带箭头的线指明父子节点之间的关系，同时以蓝色圆圈表示节点。

9.4　使用 ggtree 包绘制其他树形结构

ggtree 包可用于任何层次结构数据的可视化。下面以 2014 年的 GNI（国民总收入）数据作为示例。首先要准备好边列表，也就是一个由两列数据组成，包含父子节点关系信息的数据框或矩阵。然后通过 treeio 包中的 as.phylo() 函数将边列表转换为 phylo 对象。最后使用 ggtree 包对其与相关联的数据一起进行可视化。本示例以人口为标尺对代表不同国家的圆点的大小进行了缩放，如图 9.7 所示。

```
library(treeio)
library(ggplot2)
library(ggtree)

data("GNI2014", package="treemap")
n <- GNI2014[, c(3,1)]
n[,1] <- as.character(n[,1])
n[,1] <- gsub("\\s\\(.*\\)", "", n[,1])

w <- cbind("World", as.character(unique(n[,1])))

colnames(w) <- colnames(n)
edgelist <- rbind(n, w)
```

```
y <- as.phylo(edgelist)
ggtree(y, layout='circular') %<+% GNI2014 +
    aes(color=continent) +
    geom_tippoint(aes(size=population), alpha=.6) +
    geom_tiplab(aes(label=country), offset=.1) +
    theme(plot.margin=margin(60,60,60,60))
```

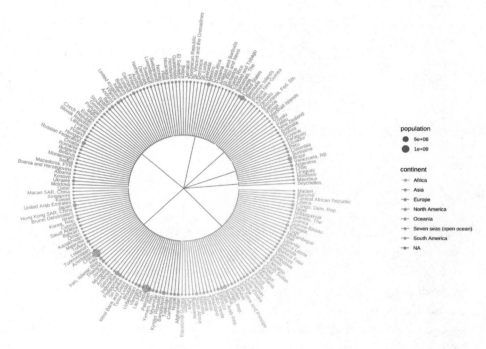

图 9.7　对任意层次结构数据进行可视化

由边连接点呈现的层次数据可以被转换为 phylo 对象，并使用 ggtree 包对其进行可视化，以此来探索它们之间的关系。

9.5　总结

　　ggtree 包支持对多种在 R 及其拓展包中定义的树对象，这也使得 ggtree 包能非常轻松地被整合到现有的各种管道中。不仅如此，ggtree 包还支持外部数据的整合，并能在树中对这些数据进行探索，这将极大地促进现有管道下游分析中的数据可视化及结果解释。最重要的是，我们对于将边列表转换为树对象的支持可以将更多的树形结构纳入 treeio 包和 ggtree 包的框架中。使得更多不同领域的树形结构，以及相关的异构数据通过 treeio 包及 ggtree 包进行整合与可视化，这将促进整合分析与比较分析，从而发现更多系统性模式与见解。

9.6 本章练习题

1. 简述 ggtree 可以衔接哪些包输出的系统发育树对象，自行构建一个系统发育树对象并对其进行可视化。

2. 尝试对 R 内置数据集 UScitiesD 进行层次聚类，将结果转换为 dendrogram 对象并使用 ggtree 对其进行可视化。

参考文献

[1] Wang L, Lam T T, Xu S, et al. treeio: an R package for phylogenetic tree input and output with richly annotated and associated data[J]. Molecular Biology and Evolution, 2020, 37(2): 599-603.

[2] Yu G, Smith D K, Zhu H, et al. ggtree: an R package for visualization and annotation of phylogenetic trees with their covariates and other associated data[J]. Methods in Ecology and Evolution, 2016, 8(1): 28-36.

[3] Yu G, Lam T T, Zhu H, et al. Two methods for mapping and visualizing associated data on phylogeny using ggtree[J]. Mol Biol Evol, 2018, 35(12): 3041-3043.

[4] Kuczynski J, Stombaugh J, Walters W A, et al. Using QIIME to analyze 16S rRNA gene sequences from microbial communities[J]. Current Protocols in Bioinformatics, 2011, 36(1): 10.7.1-10.7.20.

[5] Schloss P D, Westcott S L, Ryabin T, et al. Introducing mothur: open-source, platform-independent, community-supported software for describing and comparing microbial communities[J]. Appl Environ Microbiol, 2009, 75(23): 7537-7541.

[6] Callahan B J, McMurdie P J, Rosen M J, et al. DADA2: High-resolution sample inference from Illumina amplicon data[J]. Nat Methods, 2016, 13(7): 581-583.

[7] Kunin V H P. PyroTagger : A fast , accurate pipeline for analysis of rRNA amplicon pyrosequence data. [J]. The Open Journal, 2010: 1-8.

[8] Yu G. Using ggtree to Visualize Data on Tree-Like Structures[J]. Current Protocols in Bioinformatics, 2020, 69(1): e96.

第 *10* 章 使用 ggtreeExtra 包在环形布局上呈现数据

10.1 简介

ggtree 包[1]能帮助我们对系统发育树及其他树形结构进行可编程的可视化，也支持将进化树相关数据以多个图层的形式在进化树上直接注释，或者与树并排绘制（参见本章参考文献[2]）。虽然 ggtree 包支持以多种布局绘制进化树，但 geom_facet() 图层函数却仅适用于直角矩形布局、环形布局、椭圆布局及倾斜布局。而 ggtree 包也没有为环形布局、扇形布局或径向布局外圈数据的呈现提供直接的支持。为了解决这个问题，我们开发了 ggtreeExtra 包。它能帮助用户在环形布局树的外圈中绘制相关的图层，并将其与树结构对齐，它也兼容矩形布局的树。

10.2 基于树的结构将图与树对齐

ggtreeExtra 包中的 geom_fruit() 图层函数用于将外部图层与树结构并排对齐。与 geom_facet() 图层函数布局类似，geom_fruit() 图层函数可在内部根据树结构对输入数据进行重新排序，并使用指定的几何对象图层通过美学映射和非变量设置对数据进行可视化。可视化后生成的图形将显示在树的外圈（环形布局等）或树的右侧（矩形布局）。

geom_fruit() 图层函数被设计为能够支持 ggplot2 包及其扩展包中定义的大部分 geom 图层。图的位置（在树外圈上的位置）由 position 参数控制。position

参数需要接收一个 position 对象，其默认值为"auto"，此时 geom_fruit() 图层函数会自行推断并尝试为指定的几何对象图层确定合适的位置。在这个过程中，geom_fruit() 图层函数会自动对 geom_bar() 图层函数使用 position_stackx() 函数，对 geom_violin() 图层函数和 geom_boxplot() 图层函数使用 position_dodgex() 函数，以及对其他的几何图层（如 geom_point() 图层函数和 geom_tile() 图层函数等）使用 position_identityx() 函数，来调整它们的位置。任何带有 position 参数的几何图层都可以与 geom_fruit() 图层函数兼容，这就使在 ggtreeExtra 包中定义的位置函数能够用于调整输出图层的位置。此外，通过 axis.params 和 grid.params 参数，我们还可以分别为当前的外部图层添加坐标轴和背景网格线。

下面示例使用了 phyloseq 包中的微生物组数据，并使用箱线图来对物种丰度数据进行可视化。geom_fruit() 图层函数根据环形布局树的结构自动重新排列丰度数据，并使用特定的 geom() 图层函数（如 geom_boxplot() 图层函数）对数据进行可视化，如图 10.1 所示。本章参考文献 [2] 中的图 1 就是使用 geom_density_ridges() 图层函数和 geom_facet() 图层函数对此数据集进行了可视化。

```r
library(ggtreeExtra)
library(ggtree)
library(phyloseq)
library(dplyr)

data("GlobalPatterns")
GP <- GlobalPatterns
GP <- prune_taxa(taxa_sums(GP) > 600, GP)
sample_data(GP)$human <- get_variable(GP, "SampleType") %in%
                        c("Feces", "Skin")
mergedGP <- merge_samples(GP, "SampleType")
mergedGP <- rarefy_even_depth(mergedGP,rngseed=394582)
mergedGP <- tax_glom(mergedGP,"Order")

melt_simple <- psmelt(mergedGP) %>%
               filter(Abundance < 120) %>%
               select(OTU, val=Abundance)

p <- ggtree(mergedGP, layout="fan", open.angle=10) +
     geom_tippoint(mapping=aes(color=Phylum),
                   size=1.5,
```

```r
                        show.legend=FALSE)
p <- rotate_tree(p, -90)

p <- p +
    geom_fruit(
        data=melt_simple,
        geom=geom_boxplot,
        mapping = aes(
                    y=OTU,
                    x=val,
                    group=label,
                    fill=Phylum,
                ),
        size=.2,
        outlier.size=0.5,
        outlier.stroke=0.08,
        outlier.shape=21,
        axis.params=list(
                        axis       = "x",
                        text.size  = 1.8,
                        hjust      = 1,
                        vjust      = 0.5,
                        nbreak     = 3,
                    ),
        grid.params=list()
    )

    p <- p +
    scale_fill_discrete(
        name="Phyla",
        guide=guide_legend(keywidth=0.8, keyheight=0.8, ncol=1)
    ) +
    theme(
        legend.title=element_text(size=9),
        legend.text=element_text(size=7)
    )
p
```

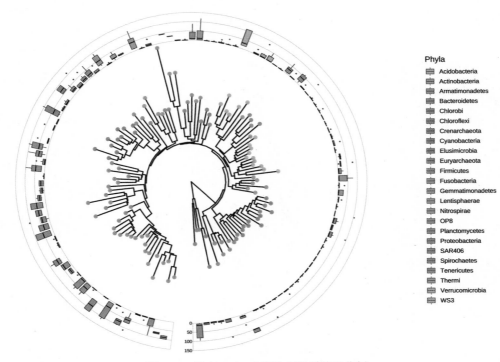

图 10.1　附有 OTU 丰度分布的系统进化树

物种丰度分布信息在与树对齐后，以箱线图的形式呈现出来。本示例使用物种所属的门信息对树上的符号点及物种丰度分布图进行着色。

10.3　在多维数据的可视化中将多个图与树对齐

我们可以为树添加多个 geom_fruit() 图层函数，对于多维数据的呈现来说，环形布局确实显得更为紧凑和高效。本示例再现了本章参考文献 [3] 中的图 2。数据由 GraPhlAn[4] 提供，其中包含不同身体部位微生物组的相对丰度。本示例向我们展示了 ggtreeExtra 包通过添加多个图层（本示例中为热图和条形图）来呈现不同类型数据的功能，如图 10.2 所示。

```
library(ggtreeExtra)
library(ggtree)
library(treeio)
library(tidytree)
library(ggstar)
```

第 10 章 使用 ggtreeExtra 包在环形布局上呈现数据

```r
library(ggplot2)
library(ggnewscale)
library(TDbook)
# 从 TDbook 中加载数据,包括 tree_hmptree、
# df_tippiont(微生物的丰度和类型)、
# df_ring_heatmap(微生物在人体不同部位的丰度)、
# 及 df_barplot_attr(流行率最高的微生物的丰度)
tree <- tree_hmptree
dat1 <- df_tippoint
dat2 <- df_ring_heatmap
dat3 <- df_barplot_attr

# 调整顺序
dat2$Sites <- factor(dat2$Sites,
                     levels=c("Stool (prevalence)", "Cheek (prevalence)",
                              "Plaque (prevalence)","Tongue (prevalence)",
                              "Nose (prevalence)", "Vagina (prevalence)",
                              "Skin (prevalence)"))
dat3$Sites <- factor(dat3$Sites,
                     levels=c("Stool (prevalence)", "Cheek (prevalence)",
                              "Plaque (prevalence)", "Tongue (prevalence)",
                              "Nose (prevalence)", "Vagina (prevalence)",
                              "Skin (prevalence)"))
# 提取进化枝标签信息
# 这是因为有些树的节点被注释至属水平,我们可以通过 ggtree 包将其突出显示
nodeids <- nodeid(tree, tree$node.label[nchar(tree$node.label)>4])
nodedf <- data.frame(node=nodeids)
nodelab <- gsub("[\\.0-9]", "", tree$node.label[nchar(tree$node.label)>4])
# 绘制进化枝标签图层,并突出显示该图层
poslist <- c(1.6, 1.4, 1.6, 0.8, 0.1, 0.25, 1.6, 1.6, 1.2, 0.4,
             1.2, 1.8, 0.3, 0.8, 0.4, 0.3, 0.4, 0.4, 0.4, 0.6,
             0.3, 0.4, 0.3)
labdf <- data.frame(node=nodeids, label=nodelab, pos=poslist)

# 环形布局树
p <- ggtree(tree, layout="fan", size=0.15, open.angle=5) +
    geom_hilight(data=nodedf, mapping=aes(node=node),
                 extendto=6.8, alpha=0.3, fill="grey", color="grey50",
                 size=0.05) +
    geom_cladelab(data=labdf,
                  mapping=aes(node=node,
                              label=label,
```

```
                                    offset.text=pos),
                    hjust=0.5,
                    angle="auto",
                    barsize=NA,
                    horizontal=FALSE,
                    fontsize=1.4,
                    fontface="italic"
                    )
p <- p %<+% dat1 + geom_star(
                        mapping=aes(fill=Phylum, starshape=Type,
                                size= Size),
                        position="identity",starstroke=0.1) +
        scale_fill_manual(values=c("#FFC125","#87CEFA","#7B68EE","#808080",
                                "#800080", "#9ACD32","#D15FEE","#FFC0CB",
                                "#EE6A50","#8DEEEE", "#006400","#800000",
                                "#B0171F","#191970"),
                        guide=guide_legend(keywidth = 0.5,
                                    keyheight = 0.5, order=1,
                                    override.aes=list(starshape =15)),
                        na.translate=FALSE)+
        scale_starshape_manual(values=c(15, 1),
                        guide=guide_legend(keywidth = 0.5,
                                    keyheight = 0.5, order=2),
                        na.translate=FALSE)+
        scale_size_continuous(range = c(1, 2.5),
                        guide = guide_legend(keywidth = 0.5,
                                    keyheight = 0.5, order=3,
                                    override.aes=list(starshape= 15)))
                                                    p <- p + new_
scale_fill() +
        geom_fruit(data=dat2, geom=geom_tile,
                mapping=aes(y=ID, x=Sites, alpha=Abundance,
                        fill= Sites),
                color = "grey50", offset = 0.04,size = 0.02)+
        scale_alpha_continuous(range=c(0, 1),
                        guide=guide_legend(keywidth = 0.3,
                                    keyheight = 0.3,
                                    order=5)) +
        geom_fruit(data=dat3, geom=geom_bar,
                mapping=aes(y=ID, x=HigherAbundance, fill=Sites),
                pwidth=0.38,
                orientation="y",
```

```
                    stat="identity",
        ) +
        scale_fill_manual(values=c("#0000FF","#FFA500","#FF0000",
                              "#800000", "#006400","#800080",
                              "#696969"),
                    guide=guide_legend(keywidth = 0.3,
                              keyheight = 0.3, order=4))+
        geom_treescale(fontsize=2, linesize=0.3, x=4.9, y=0.1) +
        theme(legend.position=c(0.93, 0.5),
            legend.background=element_rect(fill=NA),
            legend.title=element_text(size=6.5),
            legend.text=element_text(size=4.5),
            legend.spacing.y = unit(0.02, "cm"),
            )
p
```

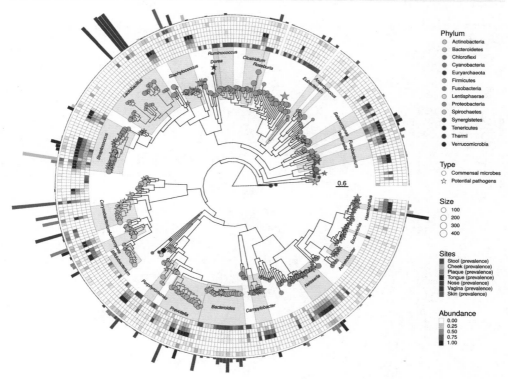

图 10.2　在系统发育树上呈现微生物组数据（丰度和位置）

进化树由符号点、突出显示的进化枝及进化枝标签注释。本示例使用了两个 geom_fruit() 图层函数对位置和丰度信息进行可视化。

本示例中叶节点的形状表示微生物的类型（共生微生物或潜在病原体），热图的透明度表示微生物的丰度，热图的颜色表示人体的不同部位，柱状图表示对应身体部位的相对丰度最高的物种的丰度。还利用节点标签中包含的分类学信息为对应的进化枝添加标签，并突出显示。

geom_fruit() 图层函数还支持矩形布局。用户可以直接为矩形布局的树添加 geom_fruit() 图层函数（如 ggtree(tree_object) + geom_fruit(...)），或者使用 layout_rectangular() 函数将环形布局树转换为矩形布局树，如图 10.3 所示。

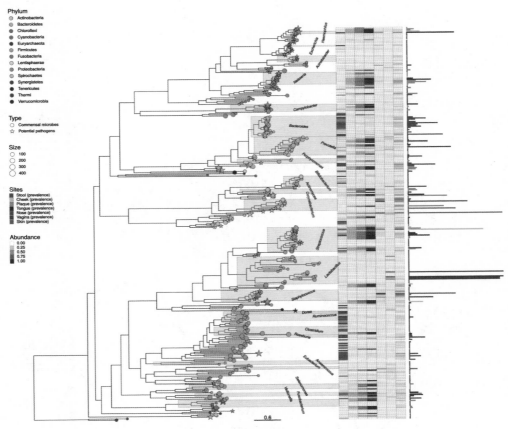

图 10.3　对矩形布局的树使用 geom_fruit() 图层函数的示例

本图是通过将图 10.2 转换为矩形布局生成的。我们同样可以将矩形布局的树转换为环形布局的树。

10.4 群体遗传学示例

ggtree 包[1] 和 ggtreeExtra 包在设计之初是作为通用工具被应用于许多研究领域的，如传染病流行病学、宏基因组学、群体遗传学、进化生物学和生态学。在此之前，我们已经介绍了在宏基因组研究中使用 ggtreeExtra 包的示例（见图 10.1 和图 10.2）。本节通过对本章参考文献 [5] 中的图 4 及本章参考文献 [6] 中的图 1 进行复现，展示 ggtreeExtra 包在群体遗传学中的应用。

```
library(ggtree)
library(ggtreeExtra)
library(ggplot2)
library(ggnewscale)
library(reshape2)
library(dplyr)
library(tidytree)
library(ggstar)
library(TDbook)
# 从 TD 包中加载 tr 数据和 dat 数据
dat <- df_Candidaauris_data
tr <- tree_Candidaauris
countries <- c("Canada", "United States",
               "Colombia", "Panama",
               "Venezuela", "France",
               "Germany", "Spain",
               "UK", "India",
               "Israel", "Pakistan",
               "Saudi Arabia", "United Arab Emirates",
               "Kenya", "South Africa",
               "Japan", "South Korea",
               "Australia")# For the tip points
dat1 <- dat %>% select(c("ID", "COUNTRY", "COUNTRY__colour"))
dat1$COUNTRY <- factor(dat1$COUNTRY, levels=countries)
COUNTRYcolors <- dat1[match(countries,dat$COUNTRY),"COUNTRY__colour"]
# 绘制热图图层
dat2 <- dat %>% select(c("ID", "FCZ", "AMB", "MCF"))
dat2 <- melt(dat2,id="ID", variable.name="Antifungal", value.name="type")
dat2$type <- paste(dat2$Antifungal, dat2$type)
```

```r
dat2$type[grepl("Not_", dat2$type)] = "Susceptible"
dat2$Antifungal <- factor(dat2$Antifungal, levels=c("FCZ", "AMB", "MCF"))
dat2$type <- factor(dat2$type,
                    levels=c("FCZ Resistant",
                             "AMB Resistant",
                             "MCF Resistant",
                             "Susceptible"))
# 绘制点图图层
dat3 <- dat %>% select(c("ID", "ERG11", "FKS1")) %>%
        melt(id="ID", variable.name="point", value.name="mutation")
dat3$mutation <- paste(dat3$point, dat3$mutation)
dat3$mutation[grepl("WT", dat3$mutation)] <- NA
dat3$mutation <- factor(dat3$mutation,
                        levels=c("ERG11 Y132F", "ERG11 K143R",
                                 "ERG11 F126L", "FKS1 S639Y/P/F"))
# 设置进化枝分组信息
dat4 <- dat %>% select(c("ID", "CLADE"))
dat4 <- aggregate(.~CLADE, dat4, FUN=paste, collapse=",")
clades <- lapply(dat4$ID, function(x){unlist(strsplit(x,split=","))})
names(clades) <- dat4$CLADE
tr <- groupOTU(tr, clades, "Clade")
Clade <- NULL
p <- ggtree(tr=tr, layout="fan", open.angle=15, size=0.2,
            aes(colour= Clade)) +
    scale_colour_manual(
        values=c("black","#69B920","#9C2E88","#F74B00","#60C3DB"),
        labels=c("","I", "II", "III", "IV"),
        guide=guide_legend(keywidth=0.5,
                           keyheight=0.5,
                           order=1,
                           override.aes=list(linetype=c("0"=NA,
                                                        "Clade1"=1,
                                                        "Clade2"=1,
                                                        "Clade3"=1,
                                                        "Clade4"=1
                                                        )
                                             )
                           )
    ) +
    new_scale_colour()
```

```r
p1 <- p %<+% dat1 +
    geom_tippoint(aes(colour=COUNTRY),
                  alpha=0) +
    geom_tiplab(aes(colour=COUNTRY),
                align=TRUE,
                linetype=3,
                size=1,
                linesize=0.2,
                show.legend=FALSE
                ) +
    scale_colour_manual(
        name="Country labels",
        values=COUNTRYcolors,
        guide=guide_legend(keywidth=0.5,
                           keyheight=0.5,
                           order=2,
                           override.aes=list(size=2,alpha=1))
    )
p2 <- p1 +
    geom_fruit(
        data=dat2,
        geom=geom_tile,
        mapping=aes(x=Antifungal, y=ID, fill=type),
        width=0.1,
        color="white",
        pwidth=0.1,
        offset=0.15
    ) +
    scale_fill_manual(
        name="Antifungal susceptibility",
        values=c("#595959", "#B30000", "#020099", "#E6E6E6"),
        na.translate=FALSE,
        guide=guide_legend(keywidth=0.5,
                           keyheight=0.5,
                           order=3
                           )
    ) +
    new_scale_fill()
p3 <- p2 +
    geom_fruit(
```

```
            data=dat3,
            geom=geom_star,
            mapping=aes(x=mutation, y=ID, fill=mutation, starshape=point),
            size=1,
            starstroke=0,
            pwidth=0.1,
            inherit.aes = FALSE,
            grid.params=list(
                        linetype=3,
                        size=0.2
                        )
    ) +
    scale_fill_manual(
        name="Point mutations",
        values=c("#329901", "#0600FF", "#FF0100", "#9900CC"),
        guide=guide_legend(keywidth=0.5, keyheight=0.5, order=4,
                        override.aes=list(
                                    starshape=c("ERG11 Y132F"=15,
                                                "ERG11 K143R"=15,
                                                "ERG11 F126L"=15,
                                                "FKS1 S639Y/P/F"=1),
                                    size=2)
                        ),
        na.translate=FALSE,
    ) +
    scale_starshape_manual(
        values=c(15, 1),
        guide="none"
    ) +
    theme(
        legend.background=element_rect(fill=NA),
        legend.title=element_text(size=7),
        legend.text=element_text(size=5.5),
        legend.spacing.y = unit(0.02, "cm")
    )
p3
```

本示例通过使用不同的颜色呈现不同的进化枝来对进化树进行注释，如图 10.4 所示。外圈的热图显示了耳念珠菌对氟康唑（FluConaZole，FCZ）、两性霉素 B（AMphotericin B，AMB）及米卡芬净（MiCaFungin，MCF）的敏感性。外圈的点图显示了羊毛甾醇 14-α- 脱甲基酶 ERG11（Y132F、K143R 和 F126L）和 β-1,3-D- 葡聚糖合酶 FKS1（S639Y/P/F）中与抗药性相关的点突变[5]。

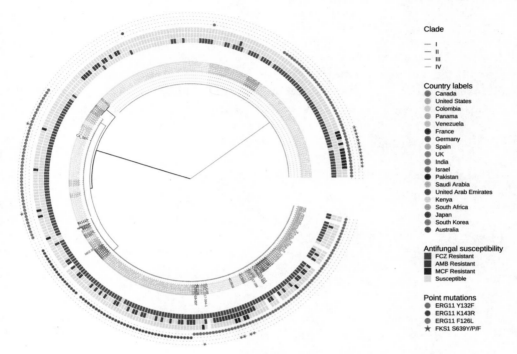

图 10.4　耳念珠菌对抗真菌药物的敏感性及其药物靶点上的点突变

```
library(ggtreeExtra)
library(ggtree)
library(ggplot2)
library(ggnewscale)
library(treeio)
library(tidytree)
library(dplyr)
library(ggstar)
library(TDbook)
# 从 TDbook 中加载 tree_NJIDqgsS 数据和 df_NJIDqgsS 数据
tr <- tree_NJIDqgsS
metada <- df_NJIDqgsS
metadata <- metada %>%
            select(c("id", "country", "country__colour",
```

```
                            "year", "year__colour", "haplotype"))
metadata$haplotype[nchar(metadata$haplotype) == 0] <- NA
countrycolors <- metada %>%
                select(c("country", "country__colour")) %>%
                distinct()
yearcolors <- metada %>%
                select(c("year", "year__colour")) %>%
                distinct()
yearcolors <- yearcolors[order(yearcolors$year, decreasing=TRUE),]
metadata$country <- factor(metadata$country, levels=countrycolors$country)
metadata$year <- factor(metadata$year, levels=yearcolors$year)
p <- ggtree(tr, layout="fan", open.angle=15, size=0.1)
p <- p %<+% metadata
p1 <-p +
    geom_tippoint(
        mapping=aes(colour=country),
        size=1.5,
        stroke=0,
        alpha=0.4
    ) +
    scale_colour_manual(
        name="Country",
        values=countrycolors$country__colour,
        guide=guide_legend(keywidth=0.3,
                            keyheight=0.3,
                            ncol=2,
                            override.aes=list(size=2,alpha=1),
                            order=1)
    ) +
    theme(
        legend.title=element_text(size=5),
        legend.text=element_text(size=4),
        legend.spacing.y = unit(0.02, "cm")
    )
p2 <-p1 +
    geom_fruit(
        geom=geom_star,
        mapping=aes(fill=haplotype),
        starshape=26,
        color=NA,
```

```
        size=2,
        starstroke=0,
        offset=0,
    ) +
    scale_fill_manual(
        name="Haplotype",
        values=c("red"),
        guide=guide_legend(
                keywidth=0.3,
                keyheight=0.3,
                order=3
            ),
        na.translate=FALSE
    )
p3 <-p2 +
    new_scale_fill() +
    geom_fruit(
        geom=geom_tile,
        mapping=aes(fill=year),
        width=0.002,
        offset=0.1
    ) +
    scale_fill_manual(
        name="Year",
        values=yearcolors$year__colour,
        guide=guide_legend(keywidth=0.3, keyheight=0.3, ncol=2,
                        order=2)
    ) +
    theme(
        legend.title=element_text(size=6),
        legend.text=element_text(size=4.5),
        legend.spacing.y = unit(0.02, "cm")
        )
p3
```

图 10.5 呈现了一个由 22145 个 SNPs[6] 推断出的伤寒沙门菌的最大似然有根树。其中叶节点的颜色表示分离株的发源地，红色符号点表示 H58 世系的单倍型，外圈热图的颜色表示菌株分离的年份 [6]。

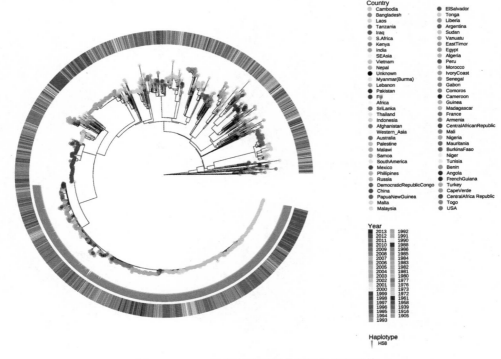

图 10.5　1832 株伤寒沙门菌分离株的种群结构

10.5　总结

与 geom_facet() 图层函数相比，ggtreeExtra 中提供的 geom_fruit() 图层函数是本章参考文献 [2] 中提出的第二种方法的更好实现。geom_facet() 图层函数和 geom_fruit() 图层函数具有相同的设计理念和相似的用户接口。它们都需要依靠其他几何对象图层来对树相关数据进行可视化。这些图层是由 ggplot2 包及其扩展包（包括 ggtree）提供的，随着 ggplot2 社区中的图层变得越来越多，geom_facet() 图层函数和 geom_fruit() 图层函数所能呈现的数据及图形的类型也会增加。

10.6　本章练习题

1. 通过 help(geom_fruit) 阅读 geom_fruit 的文档说明与示例，总结出使用

geom_fruit() 图层函数来联合外部图层的伪代码。

2. 使用以下示例数据，尝试使用 ggtreeExtra 的 geom_fruitt() 图层函数联合 ggplot2 中的 geom_col() 图层函数为系统发育树外部添加柱状图。

```
library(TDbook)
data(tree_hmptree)
data(df_barplot_attr)
```

3. 当不同图层之间有多个 fill 属性应用时，使用什么解决方案来区分不同图层的 fill 图例？

4. 在使用 geom_fruit() 图层函数时，通过哪些方式可以将外部图层的数据可视化到系统发育树上？

参考文献

[1] Yu G, Smith D K, Zhu H, et al. ggtree: an R package for visualization and annotation of phylogenetic trees with their covariates and other associated data[J]. Methods in Ecology and Evolution, 2016, 8(1): 28-36.

[2] Yu G, Lam T T, Zhu H, et al. Two methods for mapping and visualizing associated data on phylogeny using ggtree[J]. Mol Biol Evol, 2018, 35(12): 3041-3043.

[3] Morgan X C, Segata N, Huttenhower C. Biodiversity and functional genomics in the human microbiome[J]. Trends Genet, 2013, 29(1): 51-58.

[4] Asnicar F, Weingart G, Tickle T L, et al. Compact graphical representation of phylogenetic data and metadata with GraPhlAn[J]. PeerJ, 2015, 3: e1029.

[5] Chow N A, Munoz J F, Gade L, et al. Tracing the evolutionary history and global expansion of candida auris using population genomic analyses[J]. mBio, 2020, 11(2).

[6] Wong V K, Baker S, Pickard D J, et al. Phylogeographical analysis of the dominant multidrug-resistant H58 clade of Salmonella Typhi identifies inter- and intracontinental transmission events[J]. Nat Genet, 2015, 47(6): 632-639.

第 11 章　其他 ggtree 扩展包

ggtree 是用于对树结构和相关数据进行可视化的通用包。如果 ggtree 包不支持用户的某些特殊需求，则可以借助一些基于 ggtree 构建的拓展包。例如，用于对 RevBayes 输出结果进行可视化的 RevGadgets 包、对系统发育通路中固定事件进行可视化的 sitePath 包及用于对富集通路的层次结构进行可视化的 Enrichplot 包。

```
rp <- BiocManager::repositories()
db <- utils::available.packages(repo=rp)
x <- tools::package_dependencies('ggtree', db=db,
                       which = c("Depends", "Imports"),
                       reverse=TRUE)
print(x)
```

```
## $ggtree
##  [1] "enrichplot"        "ggtreeExtra"
##  [3] "LymphoSeq"         "miaViz"
##  [5] "microbiomeMarker"  "MicrobiotaProcess"
##  [7] "philr"             "singleCellTK"
##  [9] "sitePath"          "systemPipeTools"
## [11] "tanggle"           "treekoR"
```

CRAN 或 Bioconductor 共有 12 个依赖（ depend on ）或导入（ import ）ggtree 的包，GitHub 中也有一些 ggtree 的扩展包。我们在这里对一些扩展包进行简单的介绍，如 MicrobiotaProcess 包和 tangle 包。

11.1 使用 MicrobiotaProcess 包进行分类学注释

MicrobiotaProcess 包提供了一种类似 LEfSe[1] 的算法，通过比较不同类之间的分类单元丰度来发现微生物组的生物标志物。它提供了多种对分析结果进行可视化的方法。其中 ggdiffcalde 是基于 ggtree[2] 开发的。除了支持输入 diff_analysis() 的结果，它还支持输入一个包含层级关系（如分类学注释或 KEGG 注释）的数据框，以及一个包含分类单元和因子信息或 pvalue 的数据框。下面的示例演示了如何使用数据框（分析结果）对存在差异的分类树进行注释。

本示例中使用的数据框是使用 diff_analysis() 处理公共数据集[3] 得到的分析结果。我们通过设置 colors 参数将富集于相关类群中的不同特征以不同的颜色呈现。圆点的大小表示 -log10(pvalue)，即点越大颜色就越显著。从图 11.1 中我们可以发现，梭杆菌属（Fusobacterium）序列在癌中富集，而厚壁菌门（Firmicutes）、拟杆菌属（Bacteroides）和梭菌纲（Clostridiales）在肿瘤中减少。这些结果与本章参考文献 [3] 中的结论保持一致。同时，弯曲杆菌属（Campylobacter）已被证实与结直肠癌（Colorectal Cancer）有关 [4-6]。弯曲杆菌确实在肿瘤中存在富集，尽管其相对丰度要低于梭杆菌。

```
library(MicrobiotaProcess)
library(ggplot2)
library(TDbook)

# 从 TDbook 中加载 df_difftax 数据和 df_difftax_info 数据
taxa <- df_alltax_info
dt <- df_difftax

ggdiffclade(obj=taxa,
            nodedf=dt,
            factorName="DIAGNOSIS",
            layout="radial",
            skpointsize=0.6,
            cladetext=2,
            linewd=0.2,
            taxlevel=3,
            # 这个参数主要用于移除未知的分类
```

```
                reduce=TRUE) +
scale_fill_manual(values=c("#00AED7", "#009E73"))+
guides(color = guide_legend(keywidth = 0.1, keyheight = 0.6,
                            order = 3,ncol=1)) +
theme(panel.background=element_rect(fill=NA),
    legend.position="right",
    plot.margin=margin(0,0,0,0),
    legend.spacing.y=unit(0.02, "cm"),
    legend.title=element_text(size=7.5),
    legend.text=element_text(size=5.5),
    legend.box.spacing=unit(0.02,"cm")
)
```

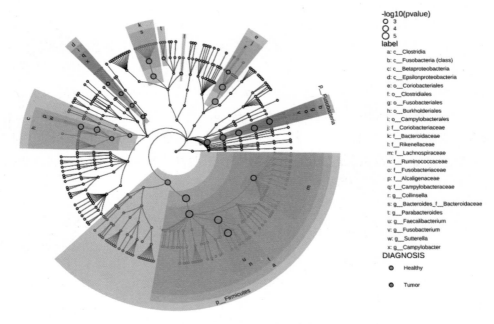

图 11.1　对存在差异的分类进化枝进行可视化

11.2　使用 tanggle 包可视化系统发育网络图

　　tanggle 包提供了一些能绘制分裂网络图（split network）的函数，通过扩展 ggtree 包[2]实现了对系统发育网络的可视化。下面使用 tanggle 包绘制系统发育网络图，如图 11.2 所示。

```
library(ggplot2)
library(ggtree)
library(tanggle)

file <- system.file("extdata/trees/woodmouse.nxs", package = 
"phangorn")
Nnet <- phangorn::read.nexus.networx(file)

ggsplitnet(Nnet) +
    geom_tiplab2(aes(color=label), hjust=-.1)+
    geom_nodepoint(color='firebrick', alpha=.4) +
    scale_color_manual(values=rainbow(15)) +
    theme(legend.position="none") +
    ggexpand(.1) + ggexpand(.1, direction=-1)
```

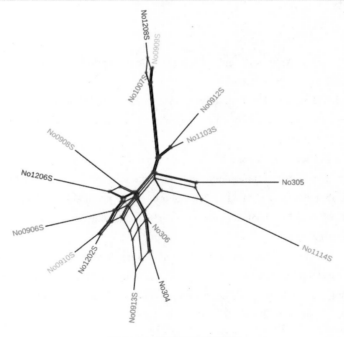

图 11.2　系统发育网络图

11.3　总结

　　ggtree 包旨在通过对图形语法的支持，帮助用户通过可视化对系统发育数据进行探索。当 ggtree 包无法满足用户的特殊需求时，可以基于 ggtree 开发相关的拓展包来实现这些缺失的功能，这是一种非常好的解决方法。我们也希望 ggtree 用户能聚集成为一个 ggtree 社区，这样就能针对各种特殊需求开发对应的功能，并在用户之间共享。这样，每个用户都将从中获益。

11.4 本章练习题

尝试使用书中的代码查看目前在 CRAN 及 Bioconductor 上依赖（depend on）或导入（import）ggtree 的包，并简述它们的主要应用领域。

参考文献

[1] Segata N, Izard J, Waldron L, et al. Metagenomic biomarker discovery and explanation[J]. Genome Biol, 2011, 12(6): R60.

[2] Yu G, Smith D K, Zhu H, et al. ggtree: an R package for visualization and annotation of phylogenetic trees with their covariates and other associated data[J]. Methods in Ecology and Evolution, 2016, 8(1): 28-36.

[3] Kostic A D, Gevers D, Pedamallu C S, et al. Genomic analysis identifies association of Fusobacterium with colorectal carcinoma[J]. Genome Res, 2012, 22(2): 292-298.

[4] Amer A, Galvin S, Healy C M, et al. The microbiome of potentially malignant oral leukoplakia exhibits enrichment for fusobacterium, leptotrichia, campylobacter, and rothia species[J]. Front Microbiol, 2017, 8: 2391.

[5] He Z, Gharaibeh R Z, Newsome R C, et al. Campylobacter jejuni promotes colorectal tumorigenesis through the action of cytolethal distending toxin[J]. Gut, 2019, 68(2): 289-300.

[6] Wu N, Yang X, Zhang R, et al. Dysbiosis signature of fecal microbiota in colorectal cancer patients[J]. Microb Ecol, 2013, 66(2): 462-470.

第 4 篇
杂项

第 12 章 ggtree 包中的实用工具

12.1 分面相关实用工具

12.1.1 facet_widths() 函数

我们经常会遇到需要调整分面面板相对宽度的情况，特别是在使用 geom_facet() 图层函数对含有关联数据的树进行可视化时，但是 ggplot2 包并不支持此功能，而 ggtree 包中 facet_widths() 函数可用于解决这个问题。对于 ggtree 对象和 ggplot 对象，我们都可以使用 facet_widths() 函数来调整分面面板的相对宽度。

```
library(ggplot2)
library(ggtree)
library(reshape2)

set.seed(123)
tree <- rtree(30)

p <- ggtree(tree, branch.length = "none") +
geom_tiplab() + theme(legend.position='none')

a <- runif(30, 0,1)
b <- 1 - a
df <- data.frame(tree$tip.label, a, b)
df <- melt(df, id = "tree.tip.label")

p2 <- p + geom_facet(panel = 'bar', data = df, geom = geom_bar,
            mapping = aes(x = value, fill = as.factor(variable)),
            orientation = 'y', width = 0.8, stat='identity') +
    xlim_tree(9)
```

```
facet_widths(p2, widths = c(1, 2))
```

facet_widths() 函数还可以使用名称向量来指定面板以调整宽度。运行以下代码的结果如图 12.1A 所示。

```
facet_widths(p2, c(Tree = .5))
```

facet_widths() 函数也适用于其他 ggplot 对象，如图 12.1B 所示。

```
p <- ggplot(iris, aes(Sepal.Width, Petal.Length)) +
  geom_point() + facet_grid(.~Species)
facet_widths(p, c(setosa = .5))
```

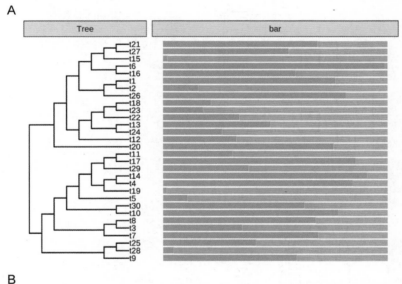

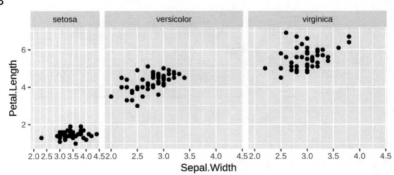

图 12.1 调整 ggplot 分面的相对宽度

facet_widths() 函数适用于 ggtree（A）对象及 ggplot（B）对象。

12.1.2 facet_labeller() 函数

facet_labeller() 函数是为更改选定面板标签而设计的,如图 12.2 所示,但它目前仅适用于 ggtree 对象(geom_facet() 图层函数的输出)。而 ggfun 包中实现了一个更为通用的版本(facet_set() 函数),可同时适用于 ggtree 对象及 ggplot 对象。

```
facet_labeller(p2, c(Tree = "phylogeny", bar = "HELLO"))
```

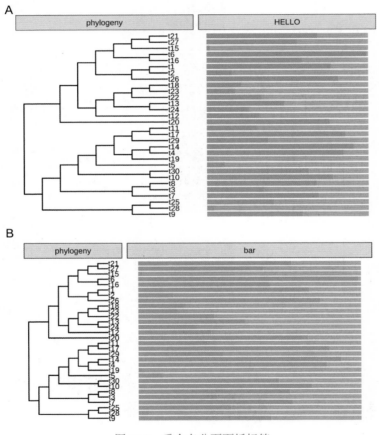

图 12.2 重命名分面面板标签

同时对多个分面面板标签重命名(A)或仅对特定的分面面板标签重命名(B)。facet_labeller() 函数可与 facet_widths() 函数联合起来使用,先重命名分面标签,再调整每个面板的相对宽度(B)。

如果想要将 facet_widths() 函数与 facet_labeller() 函数联合起来使用，则需要先调用 facet_labeller() 函数更改面板标签，再使用 facet_widths() 函数设置每个面板的相对宽度。否则，其将无法生效，这是因为 facet_widths() 函数的输出是通过 grid 对象重新绘制的。

```
facet_labeller(p2, c(Tree = "phylogeny")) %>% facet_widths(c(Tree = .4))
```

12.2　几何对象图层

在 ggplot2 包中定义的图层并不支持取子集的操作。但这个功能在对系统发育树的注释中非常有用，因为它允许我们对特定的节点进行注释（如仅为自举值大于 75 的节点添加标签）。

ggtree 包提供了一些 ggplot2 图层的修改版本，用于支持 subset 美学映射属性，其中包括 geom_segment2() 图层函数、geom_point2() 图层函数、geom_text2() 图层函数和 geom_label2() 图层函数。

这些图层函数同时适用于 ggtree 对象及 ggplot 对象，如图 12.3 所示。

```
library(ggplot2)
library(ggtree)
data(mpg)
p <- ggplot(data = mpg, mapping = aes(x = displ, y = hwy)) +
   geom_point(mapping = aes(color = class)) +
   geom_text2(aes(label=manufacturer,
             subset = hwy > 40 | displ > 6.5),
             nudge_y = 1) +
   coord_cartesian(clip = "off") +
   theme_light() +
   theme(legend.position = c(.85, .75))

p2 <- ggtree(rtree(10)) +
geom_label2(aes(subset = node <5, label = label))

plot_list(p, p2, ncol=2, tag_levels='A')
```

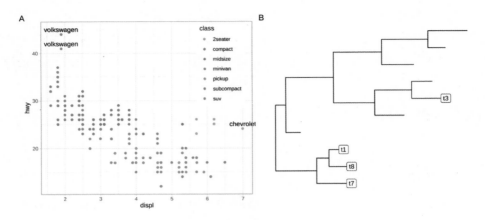

图 12.3　支持取子集的几何对象图层

我们可以在 ggplot2（A）对象和 ggtree（B）对象中使用这些图层。

12.3　布局相关工具

4.2 小节介绍了 ggtree 包支持的几种布局。除此之外，ggtree 包还额外提供了几种布局函数，可以将一种布局转换为另一种布局。需要注意的是，并非所有布局都能被转换。表 12.1 所示为布局转换函数及其说明。

表 12.1　布局转换函数及其说明

布局函数名称	说　　明
layout_circular()	将直角矩形布局转换为环形布局
layout_dendrogram()	将直角矩形布局转换为树状图布局
layout_fan()	将直角矩形布局/环形布局转换为扇形布局
layout_rectangular()	将环形布局/扇形布局转换为直角矩形布局
layout_inward_circular()	将直角矩形布局/环形布局转换为内向环形布局

```
set.seed(2019)
x <- rtree(20)
p <- ggtree(x)
p + layout_dendrogram()
ggtree(x, layout = "circular") + layout_rectangular()
p + layout_circular()
```

```
p + layout_fan(angle=90)
p + layout_inward_circular(xlim=4) + geom_tiplab(hjust=1)
```

图 12.4 所示为使用不同布局函数转换的图形效果。

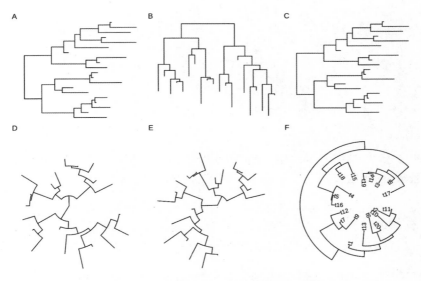

图 12.4 使用不同布局函数转换的图形效果

默认的直角矩形布局（A）；将直角矩形布局转换为树状图布局（B）；将环形布局转换为直角矩形布局（C）；将直角矩形布局转换为环形布局（D）；将直角矩形布局转换为扇形布局（E）；将直角矩形布局转换为内向环形布局（F）。

12.4 标尺相关工具

ggtree 包中的多个标尺函数用于操作 x 轴，包括 scale_x_range() 函数、xlim_tree() 函数、xlim_expand() 函数、ggexpand() 函数、hexpand() 函数和 vexpand() 函数。

12.4.1 扩大指定面板的 x 轴范围

有时用户需要为指定的面板设定 xlim（x 轴范围）（例如，在 Tree 面板上为长叶节点标签分配更多空间）。ggplot2::xlim() 函数可用于自动调节所有面板的 xlim，但是给我们带来了一些不便。而 ggtree 包中的 xlim_expand() 函数则可只用于调整用户指定面板的 xlim，并且接收 xlim 和 panel 两个参数，对所有的独立面板进行调整，如图 12.5A 所示。如果只想调整 Tree 面板中的 xlim，则可以直接使用 xlim_tree() 函数。

```
set.seed(2019-05-02)
x <- rtree(30)
p <- ggtree(x) + geom_tiplab()
d <- data.frame(label = x$tip.label,
                value = rnorm(30))
p2 <- p + geom_facet(panel = "Dot", data = d,
          geom = geom_point, mapping = aes(x = value))
p2 + xlim_tree(6) + xlim_expand(c(-10, 10), 'Dot')
```

xlim_expand() 函数也适用于 ggplot2::facet_grid() 函数。在下面示例中，使用 xlim_expand() 函数仅调整了 virginica 面板的 xlim，如图 12.5B 所示。

```
g <- ggplot(iris, aes(Sepal.Length, Sepal.Width)) +
    geom_point() + facet_grid(. ~ Species, scales = "free_x")
g + xlim_expand(c(0, 15), 'virginica')
```

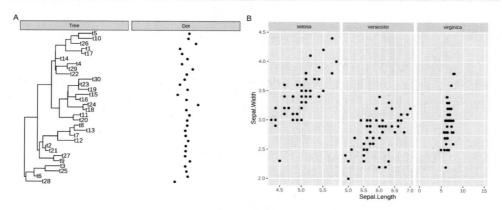

图 12.5　为用户指定的面板设置 *x* 轴范围

使用 xlim_tree() 函数设置 ggtree 包输出的 Tree 面板的 xlim（A）；使用 xlim_expand() 函数设置 ggtree 包输出的 Dot 面板的 xlim（A），以及 ggplot 包输出的 virginica 面板的 xlim（B）。

12.4.2　按一定比例扩大绘图边界

ggplot2 包并不能用于自动调整绘图边界，这使得在对长文本进行可视化时总会出现文本被截断的现象。用户需要通过 xlim() 函数（或 ylim() 函数）来手动

调节 x 轴（或 y 轴）的范围。

利用 xlim() 函数（或 ylim() 函数）确实能解决调节 x 轴（或 y 轴）范围的问题，但我们可以让这个操作变得更加简单。我们可以直接按坐标轴范围的比例放大绘图面板，这样即使不知道 x 轴、y 轴范围的确切值也能实现这个效果。

为此 hexpand() 函数可用于对 x 轴进行指定比例的延伸，并支持对延伸方向的选择（当 direction 参数的值为 1 时，向右侧延伸 x 轴；当 direction 参数的值为 -1 时，向左侧延伸 x 轴），如图 12.6 所示。ggtree 包还提供了原理类似的，可对 y 轴进行延伸的 vexpand() 函数，以及同时作用于 x 轴和 y 轴的 ggexpand() 函数。

```
x$tip.label <- paste0('to make the label longer_', x$tip.label)
p1 <- ggtree(x) + geom_tiplab() + hexpand(.3)
p2 <- ggplot(iris, aes(Sepal.Width, Petal.Width)) +
   geom_point() +
   hexpand(.2, direction = -1) +
   vexpand(.2)

plot_list(p1, p2, tag_levels="A", widths=c(.6, .4))
```

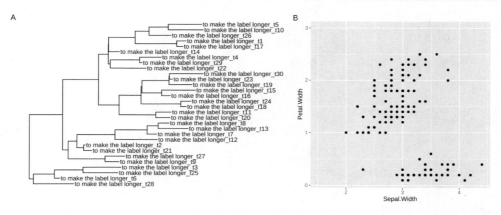

图 12.6　通过对 x 轴或 y 轴进行一定比例的延伸来拓宽绘图边界

在默认情况下，会向右侧延伸 x 轴（A）；当 direction 参数的值为 -1 时，会向左侧延伸 x 轴。向上延伸 y 轴（B）。

12.5 树数据相关工具

12.5.1 筛选树数据

ggtree 包定义了几个支持对树数据取子集的几何对象图层，但 ggplot2 包及其扩展包中许多图层并不提供此功能。为了让用户在使用这些图层时也能做到对数据进行筛选，ggtree 包中的 td_filter() 函数用于返回一个作用类似于 dplyr::filter() 的函数，并能直接传入几何对象图层中的 data 参数，以筛选 ggtree 绘图数据，如图 12.7 所示。

```
library(tidytree)

set.seed(1997)
tree <- rtree(50)
p <- ggtree(tree)
selected_nodes <- offspring(p, 67)$node
p + geom_text(aes(label=label),
          data=td_filter(isTip &
                     node %in% selected_nodes),
          hjust=0) +
    geom_nodepoint(aes(subset = node ==67),
              size=5, color='blue')
```

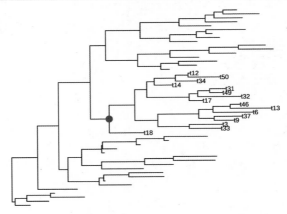

图 12.7　在几何对象图层中筛选 ggtree 绘图数据

仅对选定的叶节点添加标签（图中蓝色节点的后代）。

12.5.2 展开嵌套的树数据

ggtree 绘图数据是一个整洁数据框，其中每一行表示一个唯一的节点。如果多个值与同一个节点相关联，则这些数据将被存储为嵌套数据［存储于列表列（list-column）中］。

```
set.seed(1997)
tr <- rtree(5)
d <- data.frame(id=rep(tr$tip.label,2),
                value=abs(rnorm(10, 6, 2)),
                group=c(rep("A", 5),rep("B",5)))
require(tidyr)
d2 <- nest(d, value =value, group=group)
## d2 为嵌套数据
d2
```

```
## # A tibble: 5 × 3
##   id    value            group
##   <chr> <list>           <list>
## 1 t2    <tibble [2 × 1]> <tibble [2 × 1]>
## 2 t1    <tibble [2 × 1]> <tibble [2 × 1]>
## 3 t5    <tibble [2 × 1]> <tibble [2 × 1]>
## 4 t4    <tibble [2 × 1]> <tibble [2 × 1]>
## 5 t3    <tibble [2 × 1]> <tibble [2 × 1]>
```

我们可以通过"%<+%"操作符将嵌套数据映射到树结构上。如果几何对象图层不能对嵌套数据进行可视化，就需要在数据应用于几何对象图层前展开嵌套数据。ggtree 包中的 td_unnest() 函数的功能与 tidyr::unnest() 函数的功能类似，可用于展开嵌套的 ggtree 绘图数据，如图 12.8A 所示。

所有的树数据相关工具都提供一个".f"参数，用于传入一个可对数据进行预处理的函数。这使得我们可以组合多个树数据工具，如图 12.8B 所示。

```
p <- ggtree(tr) %<+% d2
p2 <- p +
    geom_point(aes(x, y, size= value, colour=group),
            data = td_unnest(c(value, group)), alpha=.4) +
scale_size(range=c(3,10), limits=c(3, 10))
```

```
p3 <- p +
    geom_point(aes(x, y, size= value, colour=group),
               data = td_unnest(c(value, group),
                                .f = td_filter(isTip & node==4)),
               alpha=.4) +
scale_size(range=c(3,10), limits=c(3, 10))

plot_list(p2, p3, tag_levels = 'A')
```

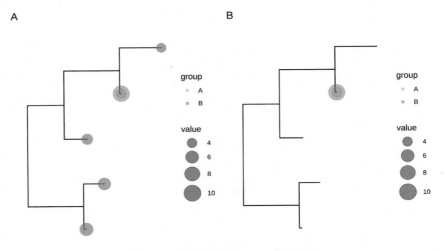

图 12.8　展开嵌套的 ggtree 绘图数据

列表列可以通过 td_unnest() 函数展开，并在叶节点处将其中的数据以两个圆点的形式同时绘制出来（A）。我们可以组合使用不同的树数据相关工具（如先通过 td_filter() 函数筛选数据，再通过 td_unnest() 函数将其展开）（B）。

12.6　树相关工具

12.6.1　提取叶节点顺序

在生成复合图（参见 7.5 小节）时，用户需要在绘制与树相关联的图之前手动对其中的数据进行重新排序，使数据顺序与利用 ggtree() 函数绘制的树结构显示的叶节点顺序保持一致。为此，ggtree 包提供了 get_taxa_name() 函数，通过该函数能提取一个叶节点向量，且该向量的数据顺序与使用 ggtree() 函数绘制的树结构显示的数据顺序保持一致。下面示例演示了 get_taxa_name() 函数的功能，如图 12.9 所示。

```
set.seed(123)
tree <- rtree(10)
p <- ggtree(tree) + geom_tiplab() +
geom_hilight(node = 12, extendto = 2.5)

x <- paste("Taxa order:",
           paste0(get_taxa_name(p), collapse=', '))
p + labs(title=x)
```

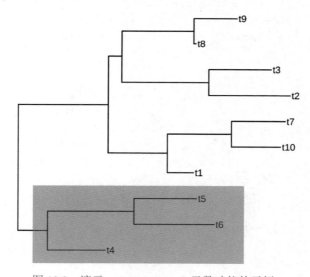

图 12.9　演示 get_taxa_name() 函数功能的示例

利用 get_taxa_name() 函数将图 12.9 中显示的树结构返回以此结构排序的叶节点标签向量。

```
get_taxa_name(p)
```

```
##   [1] "t9"  "t8"  "t3"  "t2"  "t7"  "t10" "t1"  "t5"
##   [9] "t6"  "t4"
```

如果用户指定一个节点，则可利用 get_taxa_name() 函数提取所选进化枝对应的叶节点顺序（图 12.9 中突出显示的区域）。

```
get_taxa_name(p, node = 12)
```

```
## [1] "t5" "t6" "t4"
```

12.6.2 在分类单元标签前添加填充字符

我们可以使用label_pad()函数在分类单元标签前添加填充字符（默认为"·"）。

```
set.seed(2015-12-21)
tree <- rtree(5)
tree$tip.label[2] <- "long string for test"

d <- data.frame(label = tree$tip.label,
                newlabel = label_pad(tree$tip.label),
                newlabel2 = label_pad(tree$tip.label, pad = " "))
print(d)
```

```
##                  label                 newlabel
## 1                   t1 ·················t1
## 2 long string for test long string for test
## 3                   t2 ·················t2
## 4                   t4 ·················t4
## 5                   t3 ·················t3
##              newlabel2
## 1                   t1
## 2 long string for test
## 3                   t2
## 4                   t4
## 5                   t3
```

这个功能在将叶节点标签于末端对齐时非常有用，如图12.10所示。需要注意的是，这时我们应该使用等宽字体以确保图中显示的标签长度相同。

```
p <- ggtree(tree) %<+% d + xlim(NA, 5)
p1 <- p + geom_tiplab(aes(label=newlabel),
                align=TRUE, family='mono',
                linetype = "dotted", linesize = .7)
p2 <- p + geom_tiplab(aes(label=newlabel2),
                align=TRUE, family='mono',
                linetype = NULL, offset=-.5) + xlim(NA, 5)
```

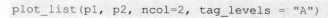

```
plot_list(p1, p2, ncol=2, tag_levels = "A")
```

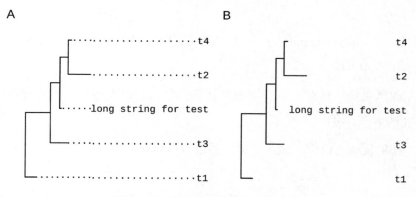

图 12.10　将叶节点标签于末端对齐

带虚线的效果（A）和不带虚线的效果（B）。

12.7　交互式 ggtree 注释

ggtree 包中的 identify() 函数提供了对进化树进行交互式的注释或操作的支持。用户可以通过单击一个节点来对对应进化枝进行突出显示、添加标签或旋转操作。用户还可以使用 plotly 包将一个 ggtree 对象转换为一个 plotly 对象，以便快速创建交互式系统发育树。

在 Youtube 和优酷上可以找到使用 identify() 函数进行交互式操作系统发育树的视频[1]。

12.8　本章练习题

1. 如何改变分面面板的尺寸及标签，当两者联用时需要注意什么？
2. 使用 iris 内置数据集绘出 Petal.Length>6 和 Sepal.Width<3 的子集的点图。
3. 如何将 ggtree(x) 对象转换为环形布局、向内环形布局、扇形布局？

---

[1] 相关交互式操作（突出显示进化枝/为进化枝添加标签/翻转进化枝）视频的网址请参见"外链资源"文档中第 12 章第 1 条

4. 使用 mpg 数据集，并基于 facet_grid(~class) 进行分面，将 2seater 面板的 xlim 修改为 1 ~ 10。

5. 输入 help(ggexpand)，阅读相关帮助文档后回答 ggexpand() 是做什么的？它又是如何工作的？

6. 随机生成一棵树，并提取出其任意一个内部节点进化支的有序叶节点顺序（与 ggtree 生成的图一致）。

7. 如何生成占位符，使得叶节点标签可以末端对齐？

第 13 章 可重复示例图库

13.1 绘制系统发育树与核苷酸序列之间的距离

本示例重现了本章参考文献 [1] 中的图 1。我们从 HPV58 树的叶节点标签中提取出了序列索引号（accession number），并计算出两两核苷酸序列之间的距离，同时将距离矩阵可视化为点线图。如图 13.1 所示，本示例演示了 ggtree 包在特定面板中添加多个图层的功能。使用 geom_facet() 图层函数可将序列距离以点图的形式呈现，并在同一个面板上添加一层线图。另外，使用 geom_facet() 图层函数在另一个面板上绘制的进化树也可以通过多层注释图层（进化枝标签、自举值支持度等）实现进化树的完全注释。本示例中源代码是从本章参考文献 [2] 的补充文件中修改的。

```
library(TDbook)
library(tibble)
library(tidyr)
library(Biostrings)
library(treeio)
library(ggplot2)
library(ggtree)

# 从 TDbook 包中加载数据
tree <- tree_HPV58

clade <- c(A3 = 92, A1 = 94, A2 = 108, B1 = 156,
           B2 = 159, C = 163, D1 = 173, D2 = 176)
tree <- groupClade(tree, clade)
cols <- c(A1 = "#EC762F", A2 = "#CA6629", A3 = "#894418", B1 = "#0923FA",
         B2 = "#020D87", C = "#000000", D1 = "#9ACD32",D2 = "#08630A")

## （可选）对树进行可视化，同时添加叶节点标签及标尺
```

```r
p <- ggtree(tree, aes(color = group), ladderize = FALSE) %>%
    rotate(rootnode(tree)) +
    geom_tiplab(aes(label = paste0("italic('", label, "')")),
                parse = TRUE, size = 2.5) +
    geom_treescale(x = 0, y = 1, width = 0.002) +
    scale_color_manual(values = c(cols, "black"),
                       na.value = "black", name = "Lineage",
                       breaks = c("A1", "A2", "A3", "B1", "B2", "C", "D1", "D2")) +
    guides(color = guide_legend(override.aes = list(size = 5, shape = 15))) +
    theme_tree2(legend.position = c(.1, .88))
## (可选) 添加单系群 (A、C 和 D) 标签及旁系群 (B) 标签
dat <- tibble(node = c(94, 108, 131, 92, 156, 159, 163, 173, 176,172),
              name = c("A1", "A2", "A3", "A", "B1",
                       "B2", "C", "D1", "D2", "D"),
              offset = c(0.003, 0.003, 0.003, 0.00315, 0.003,
                         0.003, 0.0031, 0.003, 0.003, 0.00315),
              offset.text = c(-.001, -.001, -.001, 0.0002, -.001,
                              -.001, 0.0002, -.001, -.001, 0.0002),
              barsize = c(1.2, 1.2, 1.2, 2, 1.2, 1.2, 3.2, 1.2, 1.2, 2),
              extend = list(c(0, 0.5), 0.5, c(0.5, 0), 0, c(0, 0.5),
                            c(0.5, 0), 0, c(0, 0.5), c(0.5, 0), 0)
) %>%
    dplyr::group_split(barsize)

p <- p +
    geom_cladelab(
        data = dat[[1]],
        mapping = aes(
            node = node,
            label = name,
            color = group,
            offset = offset,
            offset.text = offset.text,
            extend = extend
        ),
        barsize = 1.2,
        fontface = 3,
        align = TRUE
    ) +
    geom_cladelab(
        data = dat[[2]],
        mapping = aes(
```

```
            node = node,
            label = name,
            offset = offset,
            offset.text =offset.text,
            extend = extend
        ),
        barcolor = "darkgrey",
        textcolor = "darkgrey",
        barsize = 2,
        fontsize = 5,
        fontface = 3,
        align = TRUE
    ) +
    geom_cladelab(
        data = dat[[3]],
        mapping = aes(
            node = node,
            label = name,
            offset = offset,
            offset.text = offset.text,
            extend = extend
        ),
        barcolor = "darkgrey",
        textcolor = "darkgrey",
        barsize = 3.2,
        fontsize = 5,
        fontface = 3,
        align = TRUE
    ) +
    geom_strip(65, 71, "italic(B)", color = "darkgrey",
              offset = 0.00315, align = TRUE, offset.text = 0.0002,
              barsize = 2, fontsize = 5, parse = TRUE)
## （可选）显示支持度
p <- p + geom_nodelab(aes(subset = (node == 92), label = "*"),
                    color = "black", nudge_x = -.001, nudge_y = 1) +
    geom_nodelab(aes(subset = (node == 155), label = "*"),
                    color = "black", nudge_x = -.0003, nudge_y = -1) +
    geom_nodelab(aes(subset = (node == 158), label = "95/92/1.00"),
                    color = "black", nudge_x = -0.0001,
                    nudge_y = -1, hjust = 1) +
    geom_nodelab(aes(subset = (node == 162), label = "98/97/1.00"),
```

```
                        color = "black", nudge_x = -0.0001,
                        nudge_y = -1, hjust = 1) +
    geom_nodelab(aes(subset = (node == 172), label = "*"),
                        color = "black", nudge_x = -.0003, nudge_y = -1)
```

```
## 从叶节点标签中提取序列索引号
tl <- tree$tip.label
acc <- sub("\\w+\\|", "", tl)
names(tl) <- acc

## 将 GeneBank 中的序列直接读入 R 中，并将其转换为 DNAStringSet 对象
tipseq <- ape::read.GenBank(acc) %>% as.character %>%
    lapply(., paste0, collapse = "") %>% unlist %>%
DNAStringSet
## 使用 muscle() 函数进行序列比对
tipseq_aln <- muscle::muscle(tipseq)
tipseq_aln <- DNAStringSet(tipseq_aln)


## 计算两两序列之间的距离（Hamming Distance）
tipseq_dist <- stringDist(tipseq_aln, method = "hamming")

## 计算差异的百分比
tipseq_d <- as.matrix(tipseq_dist) / width(tipseq_aln[1]) * 100

## 将矩阵转换为整洁数据框，以在 facet_plot() 函数中使用
dd <- as_tibble(tipseq_d)
dd$seq1 <- rownames(tipseq_d)
td <- gather(dd, seq2, dist, -seq1)
td$seq1 <- tl[td$seq1]
td$seq2 <- tl[td$seq2]

g <- p$data$group
names(g) <- p$data$label
td$clade <- g[td$seq2]

## 使用点图和线图的形式对序列差异进行可视化，
## 并使用 facet_plot() 函数对序列差异图与树进行对齐
p2 <- p + geom_facet(panel = "Sequence Distance",
            data = td, geom = geom_point, alpha = .6,
            mapping = aes(x = dist, color = clade, shape = clade)) +
    geom_facet(panel = "Sequence Distance",
```

```
            data = td, geom = geom_path, alpha = .6,
            mapping=aes(x = dist, group = seq2, color = clade)) +
  scale_shape_manual(values = 1:8, guide = FALSE)

print(p2)
```

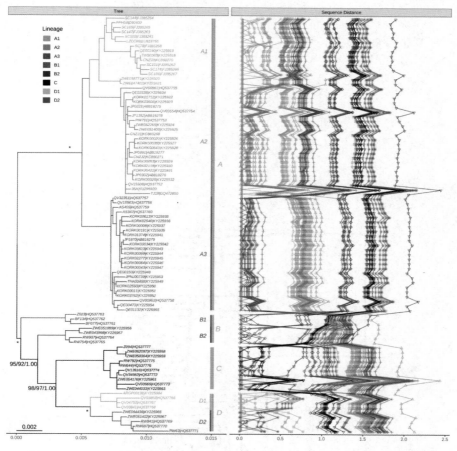

图 13.1　HPV58 完整基因组的系统发育树，附加两两序列之间距离的点线图

13.2　以不同的符号点呈现自举值

我们可以通过将自举值划为不同的区间，来表示对该进化枝的支持度是高、中还是低，并将这些区间作为分类变量，以不同颜色或形状的符号点来表示对应节点的自举值属于哪一个区间，如图 13.2 所示。

```
library(treeio)
library(ggplot2)
library(ggtree)
library(TDbook)

tree <- read.newick(text=text_RMI_tree, node.label = "support")
root <- rootnode(tree)
ggtree(tree, color="black", size=1.5, linetype=1, right=TRUE) +
    geom_tiplab(size=4.5, hjust = -0.060, fontface="bold") +
xlim(0, 0.09) +
    geom_point2(aes(subset=!isTip & node != root,
                    fill=cut(support, c(0, 700, 900, 1000))),
                    shape=21, size=4) +
    theme_tree(legend.position=c(0.2, 0.2)) +
    scale_fill_manual(values=c("white", "grey", "black"), guide='legend',
                    name='Bootstrap Percentage(BP)',
                    breaks=c('(900,1e+03]', '(700,900]', '(0,700]'),
                    labels=expression(BP>=90,70 <= BP * " " < 90", BP < 70))
```

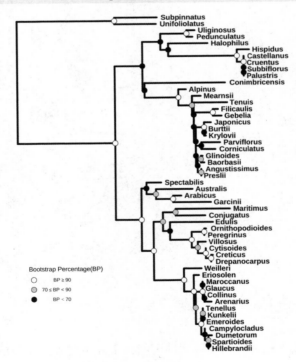

图 13.2 划分自举值

本示例将自举值分成 3 个区间，并依此对圆点进行着色。

13.3 突出显示不同分组

本示例对本章参考文献 [3] 中的图 1 进行了复现。我们先使用 groupOTU() 函数添加鸡 CTLDcps 的分组信息，并根据这个分组信息定义分支的线条类型及颜色；再使用 geom_hilight() 图层函数以不同的背景颜色突出显示两个分组的 CTLDcps（第二组为红色，第五组为绿色），同时使用 geom_cladelab() 图层函数为第五组与 A、B 亚组添加禽类特异性扩展（Avian-specific expansion）的标签，如图 13.3 所示。

```
library(TDbook)
mytree <- tree_treenwk_30.4.19

# 定义节点属性，以便稍后进行着色
tiplab <- mytree$tip.label
cls <- tiplab[grep("^ch", tiplab)]
labeltree <- groupOTU(mytree, cls)

p <- ggtree(labeltree, aes(color=group, linetype=group), layout="circular") +
    scale_color_manual(values = c("#efad29", "#63bbd4")) +
    geom_nodepoint(color="black", size=0.1) +
    geom_tiplab(size=2, color="black")

p2 <- flip(p, 136, 110) %>%
    flip(141, 145) %>%
    rotate(141) %>%
    rotate(142) %>%
    rotate(160) %>%
    rotate(164) %>%
    rotate(131)

### 对第二组和第五组进行着色
 dat <- data.frame(
        node = c(110, 88, 156,136),
        fill = c("#229f8a", "#229f8a", "#229f8a", "#f9311f")
    )
p3 <- p2 +
    geom_hilight(
        data = dat,
```

```
        mapping = aes(
            node = node,
            fill = I(fill)
        ),
        alpha = 0.2,
        extendto = 1.4
    )

### 为禽类特异性扩展部分添加标签
p4 <- p3 +
    geom_cladelab(
        node = 113,
        label = "Avian-specific expansion",
        align = TRUE,
        angle = -35,
        offset.text = 0.05,
        hjust = "center",
        fontsize = 2,
        offset = .2,
        barsize = .2
    )

### 添加 > 50 的自举值
p5 <- p4 +
    geom_nodelab(
        mapping = aes(
            x = branch,
            label = label,
            subset = !is.na(as.numeric(label)) & as.numeric(label) > 50
        ),
        size = 2,
        color = "black",
        nudge_y = 0.6
    )

### 为亚组添加标签
p6 <- p5 +
    geom_cladelab(
        data = data.frame(
            node = c(114, 121),
            name = c("Subgroup A", "Subgroup B")
        ),
```

```
        mapping = aes(
            node = node,
            label = name
        ),
        align = TRUE,
        offset = .05,
        offset.text = .03,
        hjust = "center",
        barsize = .2,
        fontsize = 2,
        angle = "auto",
        horizontal = FALSE
    ) +
    theme(
        legend.position = "none",
        plot.margin = grid::unit(c(-15, -15, -15, -15), "mm")
    )
print(p6)
```

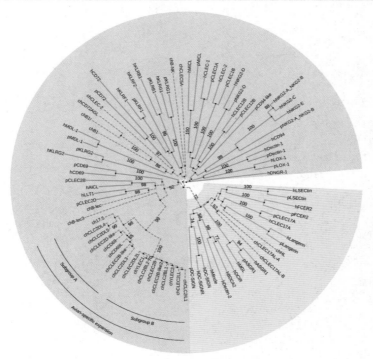

图 13.3　CTLDcps 的系统发育树

使用不同的背景颜色、线条类型及进化枝标签区分不同的分组。

13.4　含有基因组位点结构信息的系统发育树

　　ggtree 包中定义的 geom_motif() 是由封装 gggenes::geom_gene_arrow() 函数而成的图层函数。geom_motif() 图层函数可以将基因组结构根据某个指定的基因对齐（通过 on 参数），还可以通过 label 参数为所选基因添加标签。下面示例使用了 gggenes 包提供的 example_genes 数据集。由于该数据集仅提供了一组基因的基因组坐标信息，因此需要先构建基因组的系统发育树。我们基于基因组中基因交集的比例计算出杰卡德相似度(Jaccard similarity)，并以此计算出基因组距离。本示例先使用 bionj 算法来构建进化树，再使用 geom_facet() 图层函数来对含有基因组结构信息的进化树进行可视化，如图 13.4 所示。

```
library(dplyr)
library(ggplot2)
library(gggenes)
library(ggtree)

get_genes <- function(data, genome) {
filter(data, molecule == genome) %>% pull(gene)
}

g <- unique(example_genes[,1])
n <- length(g)
d <- matrix(nrow = n, ncol = n)
rownames(d) <- colnames(d) <- g
genes <- lapply(g, get_genes, data = example_genes)

for (i in 1:n) {
    for (j in 1:i) {
        jaccard_sim <- length(intersect(genes[[i]], genes[[j]])) /
                       length(union(genes[[i]], genes[[j]]))
        d[j, i] <- d[i, j] <- 1 - jaccard_sim
    }
}

tree <- ape::bionj(d)

p <- ggtree(tree, branch.length='none') +
```

```
geom_tiplab() + xlim_tree(5.5) +
geom_facet(mapping = aes(xmin = start, xmax = end, fill = gene),
           data = example_genes, geom = geom_motif, panel = 'Alignment',
           on = 'genE', label = 'gene', align = 'left') +
scale_fill_brewer(palette = "Set3") +
scale_x_continuous(expand=c(0,0)) +
theme(strip.text=element_blank(),
    panel.spacing=unit(0, 'cm'))

facet_widths(p, widths=c(1,2))
```

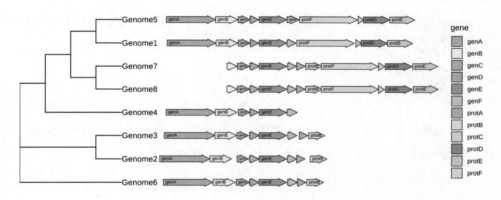

图 13.4 附有基因组特征信息的系统发育树

参考文献

[1] Chen Z, Ho W, Boon S S, et al. Ancient evolution and dispersion of human papillomavirus 58 variants[J]. J Virol, 2017, 91(21).

[2] Yu G, Lam T T, Zhu H, et al. Two methods for mapping and visualizing associated data on phylogeny using ggtree[J]. Mol Biol Evol, 2018, 35(12): 3041-3043.

[3] Larsen F T, Bed'Hom B, Guldbrandtsen B, et al. Identification and tissue-expression profiling of novel chicken c-type lectin-like domain containing proteins as potential targets for carbohydrate-based vaccine strategies[J]. Mol Immunol, 2019, 114: 216-225.

附录 A 常见问题

ggtree 邮件列表[1]是一个寻求帮助的工具。用户只需要在其中提供一个可重复的示例来说明遇到的问题，便能得到其他人的帮助。

A.1 安装相关问题

由于 ggtree 包是在 Bioconductor 项目中发布的，因此我们需要使用 BiocManager 来安装 ggtree 包。

```
## 需要先安装 BiocManager
## install.packages("BiocManager")
library(BiocManager)
install("ggtree")
```

Bioconductor 通常会随着 R 语言一起更新。如果想要安装最新版本的 Bioconductor 软件包（包括 ggtree 包），则需要保证使用的是最新版本的 R 语言。需要注意的是，错误只会在当前发行版本和开发分支中被修复。如果在安装过程中发现错误，则按照指南[2]进行报告。为了使用户仅通过一步就能轻松地安装并加载多个核心包，我们创建了一个名为 treedataverse 的元包（meta-package）。用户可以通过以下命令来安装。

```
BiocManager::install("YuLab-SMU/treedataverse")
```

安装完之后，在加载此包时也会加载 treedataverse 中的核心包，包括 tidytree、treeio、ggtree 及 ggtreeExtra。

---

[1] 邮件列表网址请参见"外链资源"文档中附录 A 第 1 条
[2] 相关指南请查看"外链资源"文档中附录 A 第 2 条

A.2　R 语言相关问题

如果 R 语言的初学者想使用 ggtree 包进行树的可视化，则需要先学习一些 R 语言及 ggplot2 包的基本知识。

一个常见的问题就是用户会直接复制粘贴命令，而根本不会查看其中函数的行为。本书中的 system.file() 函数主要用于找到在包中打包好的一些示例文件。

```
system.file                package:base                R Documentation

Find Names of R System Files

Description:

    Finds the full file names of files in packages etc.

Usage:

    system.file(..., package = "base", lib.loc = NULL,
                mustWork = FALSE)
```

对于那些想要使用自己文件的用户，请直接使用文件的绝对路径或相对路径（如 file = "your/folder/filename"）。

A.3　美学映射相关问题

A.3.1　美学映射的继承

```
ggtree(rtree(30)) + geom_point()
```

举例来说，我们可以直接通过 geom_point() 图层函数在节点处添加符号点，而不需要通过 geom_point() 图层函数中的 aes(x, y) 来提供点 x 与 y 位置的映射。这是因为它们已经在 ggtree() 函数中映射好了，并作为所有图层的全局映射。

如果我们在某个图层中提供了一个数据集，此数据集中却不包含 x、y，以及其他在 ggtree() 函数中映射好的变量，则此图层函数仍会试图映射它们。由于这些变量并不存在于数据集中，因此会收到如下报错信息。

```
Error in eval(expr, envir, enclos) : object 'x' not found
```

我们可以通过设置参数 inherit.aes = FALSE，停止从 ggtree() 函数继承美学映射来解决这个问题。

A.3.2 切忌在美学映射中使用 "$"

请千万不要这样做[1]，具体原因参考 ggplot2 book[1] 中所说的："不要在 aes() 函数中使用 "$" 引用变量（如 diamonds$carat），这样做会打破 ggplot2 包的数据容器，导致利用 ggplot2 包绘制的图不再包含所有它需要的内容，并且会在 ggplot2 包更改行顺序时产生错误，如进行分面等操作时会出现。"

A.4 文本和标签相关问题

A.4.1 叶节点标签被截断

叶节点标签被截断是由于 ggplot2 包无法根据添加的文本自动调整 xlim 参数[2] 而引起的。

```
library(ggtree)
## 从 https://support.bioconductor.org/p/72398/ 获取的示例树
tree <- read.tree(text= paste("(Organism1.006G249400.1:0.03977,",
    "(Organism2.022118m:0.01337,(Organism3.J34265.1:0.00284,",
    "Organism4.G02633.1:0.00468)0.51:0.0104):0.02469);"))
p <- ggtree(tree) + geom_tiplab()
```

在本示例中，叶节点标签被截断了，如图 A.1A 所示。这是因为点和文本分别处于数据空间和像素空间这两个不同的空间中。其解决方法有 3 种，第一种方法，用户可以通过设置 xlim 参数来为叶节点标签分配更多空间，如图 A.1B 所示。

```
p + xlim(0, 0.08)
```

第二种方法是通过设置 clip = 'off' 参数来允许将图层绘制于画布之外。在使用这种方法时，我们可能还需要设置 plot.margin 参数来延伸绘图边界，为文本留出足够的显示空间，如图 A.1C 所示。

```
p + coord_cartesian(clip = 'off') +
  theme_tree2(plot.margin=margin(6, 120, 6, 6))
```

---

[1] 相关案例的讨论请参见 "外链资源" 文档中附录 A 第 3 条
[2] 相关推文请参见附 "外链资源" 文档中附录 A 第 4 条

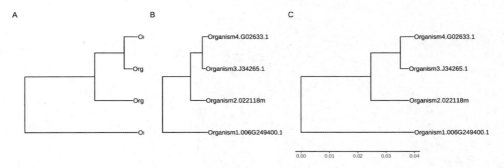

图 A.1 为被截断的标签分配更多空间

过长的叶节点标签可能会被截断（A）。一种解决方法是为绘图面板分配更多空间（B），另一种解决方法是设置允许绘图面板之外的绘制标签（C）。

第三种方法是使用 hexpand() 函数，12.4 小节有详细介绍。

对于矩形布局的树或树状图，用户还可以将叶节点标签呈现为 y 轴标签，这样无论标签有多长都不会被截断。

A.4.2 修改叶节点标签

我们可以使用 treeio::rename_taxa() 函数来修改 phylo 树对象或 treedata 树对象中的叶节点标签。

```
tree <- read.tree(text = "((A, B), (C, D));")
d <- data.frame(label = LETTERS[1:4],
                label2 = c("sunflower", "tree", "snail", "mushroom"))

## rename_taxa 默认使用第一列作为键进行匹配，第二列作为值进行替换
## rename_taxa(tree, d)
rename_taxa(tree, d, label, label2) %>% write.tree
```

```
## [1] "((sunflower,tree),(snail,mushroom));"
```

如果输入的树对象是一个 treedata 实例，则还可以使用 write.beast() 函数将含有相关数据的树导出为一个 BEAST Nexus 文件。

其实重命名进化树叶节点标签并不是一个很好的方法，这可能会导致在将原始序列比对信息映射到树时出错。我们建议还是将新的标签作为叶节点注释存储于 treedata 对象中。

```
tree2 <- full_join(tree, d, by = "label")
```

```
tree2
```

```
## 'treedata' S4 object'.
##
## ...@ phylo:
##
## Phylogenetic tree with 4 tips and 3 internal nodes.
##
## Tip labels:
##   A, B, C, D
##
## Rooted; no branch lengths.
##
## with the following features available:
##   'label2'.
```

如果只想在绘制树时展示不同或额外的信息，那么并不需要以修改叶节点标签的形式实现。我们可以通过"%<+%"操作符将修改后的标签关联到树上，再使用geom_tiplab()图层函数显示修改后的标签，如图A.2所示。

```
p <- ggtree(tree) + xlim(NA, 3)
p1 <- p + geom_tiplab()

## 该行代码功能与下一行代码功能相同
## ggtree(tree2) + geom_tiplab(aes(label = label2))
p2 <- p %<+% d + geom_tiplab(aes(label=label2))
plot_list(p1, p2, ncol=2, tag_levels = "A")
```

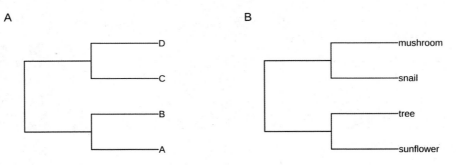

图 A.2　修改叶节点标签

原始叶节点标签（A）及修改后的叶节点标签（B）。

A.4.3 修改叶节点标签格式

如果想要修改叶节点标签的格式，则需要在 geom_text() 图层函数、geom_tiplab() 图层函数或 geom_nodelab() 图层函数中将 parse 参数设置为 TRUE，同时 label 需要是一个能被解析为 plotmath 表达式的字符串。用户还可以使用 latex2exp 包将 LaTeX 数学公式转换为 R 语言的 plotmath 表达式，或者使用 ggtext 包来渲染 Markdown 或 HTML。

本示例中的叶节点标签被分为属名、种名及地理信息等几部分。我们可以将它们以不同的格式呈现出来，以此进行区分，如图 A.3A 所示。

```
tree <- read.tree(text = "((a,(b,c)),d);")
genus <- c("Gorilla", "Pan", "Homo", "Pongo")
species <- c("gorilla", "spp.", "sapiens", "pygmaeus")
geo <- c("Africa", "Africa", "World", "Asia")
d <- data.frame(label = tree$tip.label, genus = genus,
                species = species, geo = geo)

library(glue)
d2 <- dplyr::mutate(d,
  lab = glue("italic({genus})~bolditalic({species})~({geo})"),
  color = c("#E495A5", "#ABB065", "#39BEB1", "#ACA4E2"),
  name = glue("<i style='color:{color}'>{genus} **{species}**</i> ({geo})")
)

p1 <- ggtree(tree) %<+% d2 + xlim(NA, 6) +
    geom_tiplab(aes(label=lab), parse=T)
```

使用 Markdown 或 HTML 来修改文本格式可能会更简单一点，如图 A.3B 所示。这个功能是由 ggtext 包实现的。

```
library(ggtext)

p2 <- ggtree(tree) %<+% d2 +
  geom_richtext(data=td_filter(isTip),
                aes(label=name), label.color=NA) +
  hexpand(.3)

plot_list(p1, p2, ncol=2, tag_levels = 'A')
```

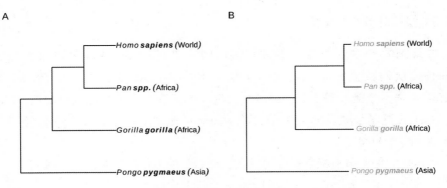

图 A.3　修改叶节点标签格式

使用 plotmath 表达式（A）和 Markdown/HTML（B）修改指定叶节点标签的格式。

A.4.4　避免文本标签重叠

用户可以通过 ggrepel 包使文本标签相互排斥而避免重叠，如图 A.4 所示。

```
library(ggrepel)
library(ggtree)
raxml_file <- system.file("extdata/RAxML",
                "RAxML_bipartitionsBranchLabels.H3",
                package="treeio")
raxml <- read.raxml(raxml_file)
ggtree(raxml) + geom_label_repel(aes(label=bootstrap,
                                fill=bootstrap)) +
  theme(legend.position = c(.1, .8)) + scale_fill_viridis_c()
```

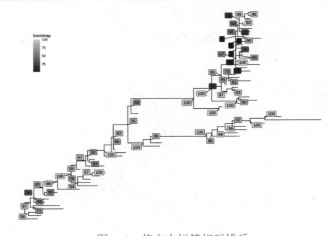

图 A.4　使文本标签相互排斥

A.4.5 Newick 格式中的自举值

在 Newick 格式中，自举值经常以内部节点标签的形式储存，如图 A.5 所示。通过下面的代码可以轻松地实现对内部节点标签的可视化。

```
geom_text2(aes(subset = !isTip, label=label))
```

如果只想展示自举值的一个子集（只展示大于 80 的自举值），就不能只是简单地使用 geom_text2(subset= (label > 80), label=label) 图层函数（或者 geom_label2() 图层函数），因为在这里 label 为字符型，其中包括了内部节点的标签（自举值）及叶节点的标签（分类单元名称）。也没有使用 geom_text 2(subset =(as . numeric(label)> 80), label=label) 图层函数，因为这样的强制转换会生成 NAs，我们还需要将 NAs 转换为逻辑值 FALSE。以下代码可以实现这样的效果。

```
nwk <- system.file("extdata/RAxML","RAxML_bipartitions.H3",
                package='treeio')
tr <- read.tree(nwk)
ggtree(tr) + geom_label2(aes(label=label,
     subset = !is.na(as.numeric(label)) & as.numeric(label) > 80))
```

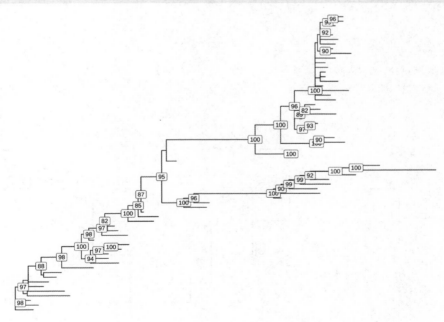

图 A.5　存储在节点标签中的自举值

这是一个十分常见的问题，我们在 treeio 包中实现了一个 read.newick() 函数。

该函数可用于将内部节点标签解析为支持度。这样，通过下面的代码就能更加轻松地展示自举值。

```
tr <- read.newick(nwk, node.label='support')
ggtree(tr) + geom_nodelab(geom='label', aes(label=support,
                                            subset=support > 80))
```

A.5 分支设置

A.5.1 绘制与 plot.phylo() 函数效果相同的树

在默认情况下，使用 ggtree() 函数会将输入的树以阶梯形呈现，使其看起来没有那么混乱。这就是使用 ggtree() 函数绘制出来的树与使用 plot.phylo() 函数绘制出来的树长得不一样的原因。如果不想将树呈现为阶梯形，则用户可以将 ladderize = FALSE 传给 ggtree() 函数，如图 A.6 所示。

```
library(ape)
library(ggtree)
set.seed(42)
x <- rtree(5)
plot(x)
ggtree(x, ladderize = FALSE) + geom_tiplab()
ggtree(x) + geom_tiplab()
```

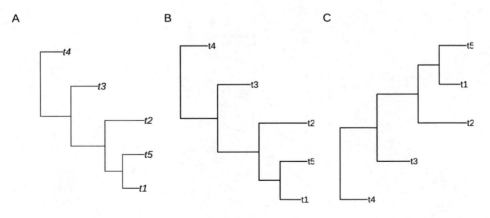

图 A.6 阶梯形树和非阶梯形树

使用 plot.phylo() 函数将树以非阶梯形呈现（A），通过设置 ladderize = FALSE 使 ggtree() 函数生成非阶梯形的树（B），使用 ggtree() 函数默认将树以阶梯形呈现（C）。

A.5.2 指定叶节点的顺序

ape 包中的 rotateConstr() 函数会根据指定的节点顺序旋转内部分支。在绘制树时，叶节点便会以这种顺序呈现（从下到上排列）。由于 ggtree() 函数在默认情况下会将输入的树以阶梯形呈现，所以我们首先设置 ladderize = FALSE，然后树就能正常地根据用户指定的节点顺序绘制，如图 A.7 所示。用户还可以通过 ggtree() 函数中的 get_taxa_name() 函数将叶节点标签的顺序提取出来。

```
y <- ape::rotateConstr(x, c('t4', 't2', 't5', 't1', 't3'))
ggtree(y, ladderize = FALSE) + geom_tiplab()
```

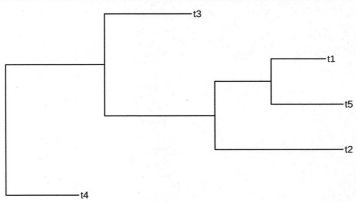

图 A.7 指定树的顺序

当设置 ladderize = FALSE 时，ggtree() 函数便能根据输入的叶节点的顺序对树进行可视化。

A.5.3 缩短外群长分支

当外群位于枝长很长的分支时，如图 A.8A 所示，我们希望将外群保留在树中，同时忽略它们的枝长，如图 A.8B 所示[1]。我们可以通过修改外群的坐标轻松地达到这个效果，如图 A.8B 所示。还有一种方法是使用 ggbreak 包[2] 对图进行截断，如图 A.8C 所示。

```
library(TDbook)
library(ggtree)

x <- tree_long_branch_example
m <- MRCA(x, 75, 76)
y <- groupClade(x, m)
```

① 相关讨论请参见"外链资源"文档中附录 A 第 5 条

```
## A
p <- p1 <- ggtree(y, aes(linetype = group)) +
    geom_tiplab(size = 2) +
    theme(legend.position = 'none')

## B
p$data[p$data$node %in% c(75, 76), "x"] <- mean(p$data$x)

## C
library(ggbreak)
p2 <- p1 + scale_x_break(c(0.03, 0.09)) + hexpand(.05)

## 对齐图
plot_list(p1, p, p2, ncol=3, tag_levels="A")
```

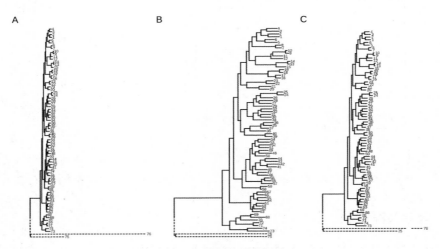

图 A.8 缩短外群的长分支

原始的树（A）；缩短外群分支长度后的树（B）；截断树图（C）。

A.5.4 为树添加新的叶节点

有时候，一些已知的分支并没有在树中出现，而我们希望将其添加到树上。还有一种常见的情况是，当有一些新测序的物种时，我们希望通过推断其在参考树上的进化位置来更新参考树。

这时，我们可以通过 phytools::bind.tip() 函数[3]将新的叶节点添加到树上。而通过 tidytree，我们也能为新添加的分支添加注释，并对其与原本的分支进行区分，展示添加分支位置的不确定性，如图 A.9 所示。

```r
library(phytools)
library(tidytree)
library(ggplot2)
library(ggtree)

set.seed(2019-11-18)
tr <- rtree(5)

tr2 <- bind.tip(tr, 'U', edge.length = 0.1, where = 7, 
position=0.15)
d <- as_tibble(tr2)
d$type <- "original"
d$type[d$label == 'U'] <- 'newly introduced'
d$sd <- NA
d$sd[parent(d, 'U')$node] <- 0.05

tr3 <- as.treedata(d)
ggtree(tr3, aes(linetype=type)) + geom_tiplab() +
  geom_errorbarh(aes(xmin=x-sd, xmax=x+sd, y = y - 0.3),
                 linetype='dashed', height=0.1) +
  scale_linetype_manual(values = c("newly introduced" = "dashed",
                                   "original" = "solid")) +
  theme(legend.position=c(.8, .2))
```

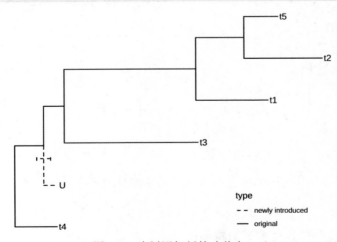

图 A.9　为树添加新的叶节点

本示例使用了不同的线条类型来区分新添加的叶节点，也添加了一个误差条以显示添加分支位置的不确定性。

A.5.5 更改任意分支的颜色或线条类型

如果想要更改任意分支的颜色或线条类型，则只需要准备一个含有分支设置变量（如哪些分支是所选的，哪些不是）的数据框。通过使用前文及附录 A 参考文献 [4] 中介绍的第一种方法，将数据映射到树的结构，便能轻松地调整任意分支的颜色或线条类型，如图 A.10 所示。

```
set.seed(123)
x <- rtree(10)
## 设置两种颜色
d <- data.frame(node=1:Nnode2(x), colour = 'black')
d[c(2,3,14,15), 2] <- "red"

## 设置多种线型
d2 <- data.frame(node=1:Nnode2(x), lty = 1)
d2[c(2,5,13, 14), 2] <- c(2, 3, 2,4)

p <- ggtree(x) + geom_label(aes(label=node))
p %<+% d %<+% d2 + aes(colour=I(colour), linetype=I(lty))
```

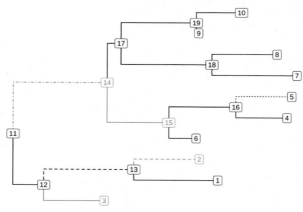

图 A.10　更改任意分支的颜色或线条类型

用户还可以使用 gginnards 包，通过直接操作图中的元素来应对一些更为复杂的场景。

A.5.6 在分支的任意位置添加符号点

如果想要在分支的任意位置添加一个符号点[①]，则使用 geom_nodepoint() 图层函

[①] 相关问题详情请参见"外链资源"文档中附录 A 第 6 条

数、geom_tippoint()图层函数或 geom_point2()图层函数（同时适用于内部及外部节点）中的 subset 美学映射筛选出需要添加符号点的分支的对应节点（所选分支的终点），并通过 x = x − offset 美学映射指定要添加的点的水平位置。其中，offset（偏移值）可以是一个绝对值（见图 A.11A），也可以是一定比例的枝长（见图 A.11B）。

```
set.seed(2020-05-20)
x <- rtree(10)
p <- ggtree(x)

p1 <- p + geom_nodepoint(aes(subset = node == 13, x = x - .1),
                         size = 5, colour = 'firebrick', shape = 21)

p2 <- p + geom_nodepoint(aes(subset = node == 13,
                             x = x - branch.length * 0.2),
                         size = 3, colour = 'firebrick') +
    geom_nodepoint(aes(subset = node == 13, x = x - branch.length * 0.8),
                   size = 5, colour = 'steelblue')
plot_list(p1, p2, ncol=2, tag_levels="A")
```

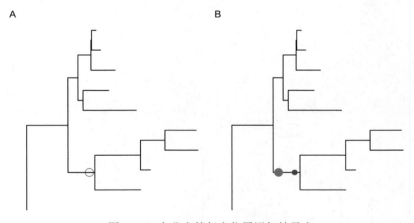

图 A.11　在分支的任意位置添加符号点

符号点的位置可以根据绝对值（A）或一定比例的枝长（B）进行调整。

A.6　为不同的分面面板设置不同的 x 轴标签

ggplot2 通常并不支持此种做法，但我们可以通过在每个面板上绘制文本标签，并将标签放置于绘图面板之外来实现，如图 A.12 所示。

```
library(ggtree)
library(ggplot2)
set.seed(2019-05-02)
x <- rtree(30)
p <- ggtree(x) + geom_tiplab()
d <- data.frame(label = x$tip.label,
                value = rnorm(30))
p2 <- p + geom_facet(panel = "Dot", data = d,
          geom = geom_point, mapping = aes(x = value))

p2 <- p2 + theme_bw() +
   xlim_tree(5) + xlim_expand(c(-5, 5), 'Dot')

#.panel 是 geom_facet() 图层函数中用于分面的内部变量
d <- data.frame(.panel = c('Tree', 'Dot'),
                lab = c("Distance", "Dot Units"),
                x=c(2.5,0), y=-2)

p2 + scale_y_continuous(limits=c(0, 31),
                        expand=c(0,0),
                        oob=function(x, ...) x) +
   geom_text(aes(label=lab), data=d) +
   coord_cartesian(clip='off')  +
   theme(plot.margin=margin(6, 6, 40, 6))
```

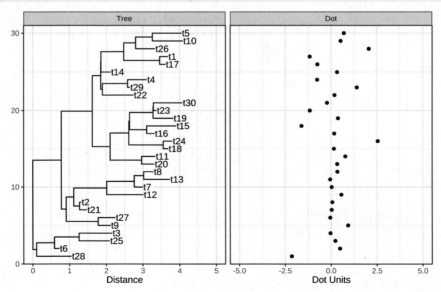

图 A.12　为不同分面面板设置不同的 *x* 轴标签

A.7 在树的底部图层绘制图形

使用 ggtree() 函数在将树结构绘制出来后，通常会在所有图层的上面继续添加后续的图层，如图 A.13 所示。

```
set.seed(1982)
x <- rtree(5)
p <- ggtree(x) + geom_hilight(node=7, alpha=1)
```

如果想要将这些图层添加到树图层的下面，则可以通过颠倒所有图层的顺序来实现。

```
p$layers <- rev(p$layers)
```

还有一种方法是使用 ggplot() 函数代替 ggtree() 函数，使用 "+geom_tree() 图层函数"将树结构的图层添加到图层栈中的合适位置。

```
ggplot(x) + geom_hilight(node=7, alpha=1) + geom_tree() + theme_tree()
```

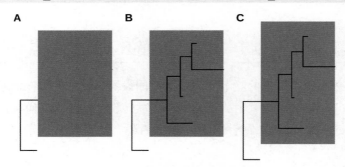

图 A.13　在树结构底层添加图层

在树结构上层添加图层（A）。颠倒 A 图中图层的顺序（B）。在树图层之下添加图层（C）。

A.8 扩大环形布局或扇形布局树的内部空间

有很多人都提过这个问题[①]，在本章参考文献 [5] 中能找到一个已发表的示例。增加环形布局树内部空间的比例可以很好地避免叶节点标签出现重叠，也能通过将所有的节点及分支向外推移而提高树的可读性。我们可以使用 xlim() 函数或 hexpand() 函数来为树分配更多空间，如图 A.14A 所示，或者为树设置一个很长的根分支，类似于 FigTree 通过设置 Root Length 参数达到的效果。

① 相关讨论请参见"外链资源"文档中附录 A 第 7 条

```
set.seed(1982)
tree <- rtree(30)
plot_list(
  ggtree(tree, layout='circular') + xlim(-10, NA),
  ggtree(tree, layout='circular') + geom_rootedge(5),
  tag_levels = "A", ncol=2
)
```

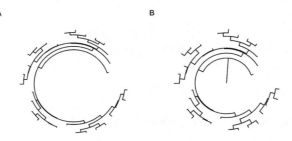

图 A.14　扩大环形布局树的内部空间

我们可以使用 xlim() 函数（A）或为树设置一个很长的根分支（B）来分配更多空间。

A.9　使用离根最远的叶节点作为时间尺度树的原点

revts() 函数用于将最近的叶节点表示的时间设置为 0，以此来翻转 x 轴。在使用 revts() 函数后，x 轴的最大值为 0，其余值为负数。我们可以通过设置 scale_x_continuous(labels=abs) 函数来使绝对值作为 x 轴标签，如图 A.15 所示。

```
tr <- rtree(10)
p <- ggtree(tr) + theme_tree2()
p2 <- revts(p) + scale_x_continuous(labels=abs)
plot_list(p, p2, ncol=2, tag_levels="A")
```

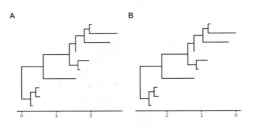

图 A.15　设置时间尺度的起点

向前：从根到叶节点（A）。

向后：从最远的叶节点到根（B）。

A.10　删除环形布局树的空白边距

在极坐标下绘制的图（如环形布局的树）经常会在图的外部生成一些多余的空白区域。如果想要使用 Rmarkdown 删除多余的空白区域，则可以通过设置 knitr 参数来完成。

```
library(knitr)
knit_hooks$set(crop = hook_pdfcrop)
opts_chunk$set(crop = TRUE)
```

或者使用命令行工具删除多余的空白区域。

```
## for pdf
pdfcrop x.pdf

## for png
convert -trim x.png x-crop.png
```

如果想要在 R 语言中删除多余的空白区域，则可以使用 magick 包来完成。

```
library(magick)
x <- image_read("x.png")
## 如果为 PDF 文件
## x <- image_read_pdf("x.pdf")
image_trim(x)
```

下面示例的原始图与修剪后的图如图 A.16 所示。

```
library(ggplot2)
library(ggtree)
library(patchwork)
library(magick)

set.seed(2021)
tr <- rtree(30)
p <- ggtree(tr, size=1, colour="purple", layout='circular')

f <- tempfile(fileext=".png")
ggsave(filename = f, plot = p, width=7, height=7)

x <- image_read(f, density=300)
y <- image_trim(x)

panel_border <- theme(panel.border=element_rect(colour='black',
```

```
                                              fill=NA, size=2))
xx <- image_ggplot(x) + panel_border
yy <- image_ggplot(y) + panel_border

plot_list(xx, yy, tag_levels = "A", ncol=2)
```

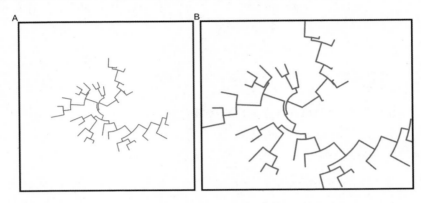

图 A.16　删除极坐标下的多余空白区域

原始图（A）与修剪后的图（B）。

A.11　编辑树图的细节

对于普通用户来说，使用 ggplot2 包或 ggtree 包来修改图的细节是一件比较困难的事。我们推荐使用 eoffice 包将 ggtree 包的输出结果导入一个 Microsoft Office 文档，并在 PowerPoint 中编辑树图。

参考文献

[1] Wickham H. ggplot2: Elegant graphics for data analysis[M]. 2016.

[2] Xu S, Chen M, Feng T, et al. Use ggbreak to effectively utilize plotting space to deal with large datasets and outliers[J]. Front Genet, 2021, 12: 774846.

[3] Revell L J. phytools: an R package for phylogenetic comparative biology (and other things)[J]. Methods in Ecology and Evolution, 2012, 3(2): 217-223.

[4] Yu G, Lam T T, Zhu H, et al. Two methods for mapping and visualizing associated data on phylogeny using ggtree[J]. Mol Biol Evol, 2018, 35(12): 3041-3043.

[5] Barton K, Hiener B, Winckelmann A, et al. Broad activation of latent HIV-1 in vivo[J]. Nat Commun, 2016, 7: 12731.

附录 B 相关工具

B.1 MircrobiotaProcess 包：将物种分类表转换为 treedata 对象

分类信息（属、科等）数据在微生物组学与生态学中的应用非常广泛。层级分类法是一种树状结构，将个体组织成子类别，并且可以转换为树对象。MircrobiotaProcess 包支持将 phyloseq 包定义的 taxonomyTable 对象转换为 treedata 对象，由此，分类层级关系便可使用 ggtree 包进行可视化，如图 B.1 所示。当分类学名称出现混杂或缺失的情况时（如不同等级的物种含有相同的分类学名称等），可以使用 taxonomyTable 对象中的 as.treedata() 函数为其自动补充上一级的分类信息。

```r
library(MicrobiotaProcess)
library(ggtree)

# 原始的 kostic2012crc 是一个 MPSE 对象
data(kostic2012crc)

taxa <- tax_table(kostic2012crc)
# 当 include.rownames = TRUE 时，
# 分类单元的行名（一般是 OTUs 或其他特征）会被作为叶节点标签
tree <- as.treedata(taxa, include.rownames=TRUE)
# 使用 mp_extract_tree() 函数提取分类单元树（treedata 对象），
# 因为分类学信息在 MPSE 类 (kostic2012crc) 中会以 treedata 对象的形式存储
# tree <- kostic2012crc %>% mp_extract_tree()

ggtree(tree, layout="circular", size=0.2) +
    geom_tiplab(size=1)
```

R 实战：系统发育树的数据集成操作及可视化

图 B.1　将 taxonomyTable 对象转换为 treedata 对象

B.2　rtol 包：Open Tree API 的 R 接口

rtol[1] 是一个能与 Open Tree of Life（简称 Open Tree）数据 APIs 交互的 R 包。我们可以先通过 rtol 包来获取系统发育树，再通过 ggtree 包对系统发育树进行可视化并探索其中的物种关系，如图 B.2 所示。

```
## 示例来源于 https://github.com/ropensci/rotl
library(rotl)
apes <- c("Pongo", "Pan", "Gorilla", "Hoolock", "Homo")
(resolved_names <- tnrs_match_names(apes))

##   search_string unique_name approximate_match ott_id
## 1         pongo       Pongo             FALSE 417949
## 2           pan         Pan             FALSE 417957
## 3       gorilla     Gorilla             FALSE 417969
## 4       hoolock     Hoolock             FALSE 712902
## 5          homo        Homo             FALSE 770309
##   is_synonym          flags number_matches
## 1      FALSE                             2
## 2      FALSE sibling_higher              2
## 3      FALSE sibling_higher              1
## 4      FALSE                             1
## 5      FALSE sibling_higher              1
```

```
tr <- tol_induced_subtree(ott_ids = ott_id(resolved_names))
ggtree(tr) + geom_tiplab() + xlim(NA, 5)
```

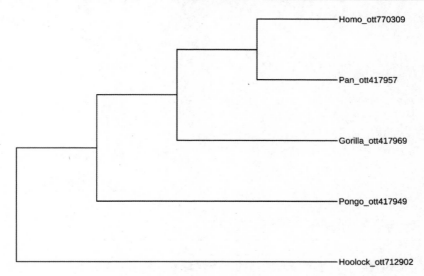

图 B.2　从完整的 Open Tree of Life 中获取一个诱导子树

B.3　将 ggtree 对象转换为 plotly 对象

通过 plotly 包来使用 ggtree 包，能够快速地构建一个交互式系统发育树。通过 ggplotly() 函数能将 ggtree 对象转换为 plotly 对象。需要注意的是，使用 ggtree 包中的 identify() 函数可以实现系统发育树的交互式操作。

```
# 示例来自 https://twitter.com/drandersgs/status/965996335882059776

# 加载库
library(ape)
library(ggtree)
library(plotly)
# 构建系统发育树
n_samples <- 20
n_grp <- 4
tree <- ape::rtree(n = n_samples)
# 创建一些元数据
id <- tree$tip.label
```

```
set.seed(42)
grp <- sample(LETTERS[1:n_grp], size = n_samples, replace = T)
dat <- tibble::tibble(id = id,
                      grp = grp)
# 绘制系统发育树
p1 <- ggtree(tree)
metat <- p1$data %>%
  dplyr::inner_join(dat, c('label' = 'id'))
p2 <- p1 +
  geom_point(data = metat,
             aes(x = x,
                 y = y,
                 colour = grp,
                 label = id))
plotly::ggplotly(p2)
```

B.4 绘制漫画风格的系统发育树（类似 xkcd）

运行下面代码可以绘制漫画风格的系统发育树，如图 B.3 所示。

```
library(htmltools)
library(XML)
library(gridSVG)
library(ggplot2)
library(ggtree)
library(comicR)

p <- ggtree(rtree(30), layout="circular") +
    geom_tiplab(aes(label=label), color="purple")
print(p)
svg <- grid.export(name="", res=100)$svg

tagList(
    tags$div(
            id = "ggtree_comic",
            tags$style("#ggtree_comic text {font-family:Comic Sans MS;}"),
            HTML(saveXML(svg)),
            comicR("#ggtree_comic", ff=5)
        )
) # %>% html_print
```

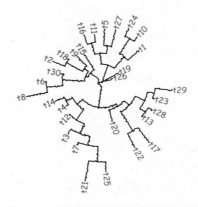

图 B.3　绘制漫画风格的系统发育树

B.5　绘制 ASCII Art 形式的有根树

```
library(data.tree)
tree <- rtree(10)
d <- as.data.frame(as.Node(tree))
names(d) <- NULL
print(d, row.names=FALSE)
```

```
 11
  ¦--12
  ¦   ¦--13
  ¦   ¦   ¦--t4
  ¦   ¦   °--t7
  ¦   °--14
  ¦       ¦--15
  ¦       ¦   ¦--t1
  ¦       ¦   °--16
  ¦       ¦       ¦--t6
  ¦       ¦       °--t5
  ¦       °--t8
  °--17
      ¦--t3
      °--18
          ¦--19
          ¦   ¦--t10
```

```
       --t9
°  --t2
```

通过 ASCII Art 的形式，我们可以将进化树以一种简洁的方式呈现出来。有时候，我们并不想绘制出整棵树，而只是想在 R 控制台上观看一下树的结构，ASCII Art 形式的树就能做到这点。如果树很大，那么最好还是不要这样操作。有时候，我们只想聚焦于树的特定部分及这部分的直系亲属。此时我们可以先使用 treeio::tree_subset() 函数提取出树的选定部分，再通过输出树子集的 ASCII Art，在 R 控制台上探索我们感兴趣的生物间的进化关系。

ggtree 包还支持将叶节点标签解析为表情符号来构建 phylomoji。data.tree 包及 emojifont 包支持以 ASCII 文本的形式输出 phylomoji，如图 B.4 所示。

```
library(data.tree)
library(emojifont)

tt <- '((snail,mushroom),(((sunflower,evergreen_tree),leaves),green_salad));'
tree <- read.tree(text = tt)
tree$tip.label <- emoji(tree$tip.label)
d <- as.data.frame(as.Node(tree))
names(d) <- NULL
print(d, row.names=FALSE)
```

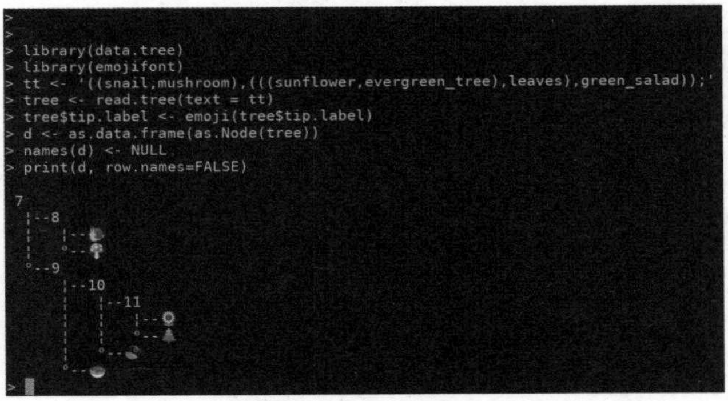

图 B.4 以 ASCII 文本的形式输出 phylomoji

还有一种输出系统发育树的 ASCII ART 的方法是，使用 devout 包中定义的 ascii() 函数绘制树，如下示例。

```
library(devout)
ascii(width=80)
ggtree(rtree(5))
invisible(dev.off()
```

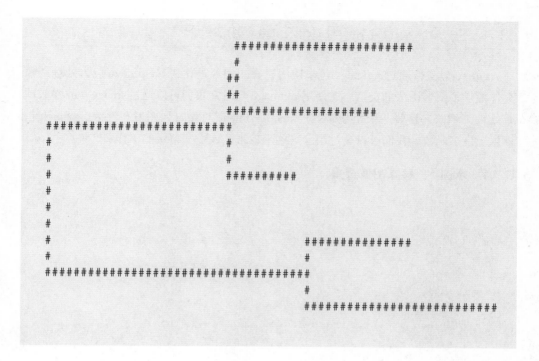

B.6 放大树的选定部分

除了使用 viewClade() 函数,用户还可以通过 ggforce 包来放大选定的进化枝,如图 B.5 所示。

```
set.seed(2019-08-05)
x <- rtree(30)
nn <- tidytree::offspring(x, 43, self_include=TRUE)
ggtree(x) + ggforce::facet_zoom(xy = node %in% nn)
```

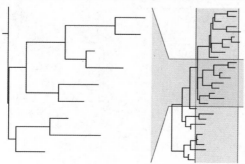

图 B.5 放大选定的进化支

B.7 在 ggtree 包中使用 ggimage 包的提示

ggtree 包支持通过 ggimage 包来使用轮廓图注释进化树。ggimage 包为处理图像文件提供了图形语法的支持。在使用 ggimage 包时，我们可以通过 image_fun 参数即时处理图像，并接收一个函数来处理 magick-image 对象，如图 B.6 所示。magick 包提供了许多函数，我们可以通过组合这些函数来完成一些特定的工作。

B.7.1 示例 1：移除图像背景

```r
library(ggimage)

imgdir <- system.file("extdata/frogs", package = "TDbook")

set.seed(1982)
x <- rtree(5)
p <- ggtree(x) + theme_grey()
p1 <- p + geom_nodelab(image=paste0(imgdir, "/frog.jpg"),
                       geom="image", size=.12) +
    ggtitle("original image")
p2 <- p + geom_nodelab(image=paste0(imgdir, "/frog.jpg"),
          geom="image", size=.12,
          image_fun= function(.) magick::image_transparent(.,
"white")) +
    ggtitle("image with background removed")
plot_grid(p1, p2, ncol=2)
```

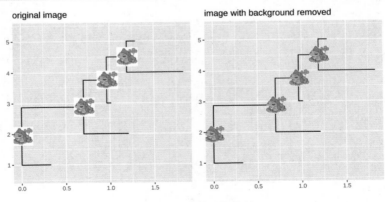

图 B.6　移除图像背景

在进化树上绘制没有移除背景的图像（A）及移除了背景的图像（B）。

B.7.2 示例 2：在背景图像上绘制树

通过 geom_bgimage() 图层函数能添加一个图像图层，并将其放在图层栈的底部。作为一个普通的图层，它不会改变输出的 ggtree 对象的结构。用户可以在此之上继续添加注释图层，如图 B.7 所示。

```
ggtree(rtree(20), size=1.5, color="white") +
  geom_bgimage('img/blackboard.jpg') +
  geom_tiplab(color="white", size=5, family='xkcd')
```

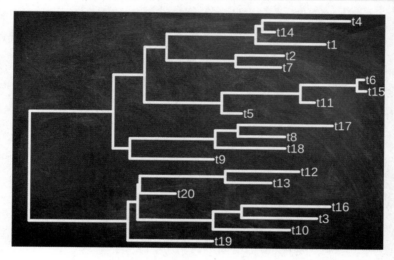

图 B.7　在背景图像上绘制树

B.8　在 Jupyter Notebook 中运行 ggtree 包

如果系统中已经安装了 Jupyter Notebook，则可以在 R 语言中通过以下命令安装 IRkernel。

```
install.packages("IRkernel")
IRkernel::installspec()
```

接下来用户就可以在 Jupyter Notebook 中使用 ggtree 包及其他 R 包，如图 B.8 所示。

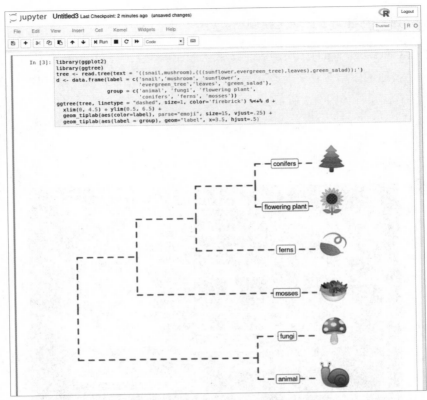

图 B.8 在 Jupyter Notebook 中运行 ggtree 包

参考文献

[1] Michonneau F, Brown J W, Winter D J. rotl: an R package to interact with the open tree of life data[J]. Methods in Ecology and Evolution, 2016, 7(12): 1476-1481.

附录 C 练习题答案

第 1 章

1. 参考答案

系统发育树（Phylogenetic Tree，简称"进化树"）是基于生物的遗传序列构建的，常用来描述生物群体之间的谱系关系。我们可以通过找出哪个病原体样本在遗传上更接近另一个样本，从而更加深入地了解平时难以观察到的流行病学联系和流行病爆发的潜在源头。我们通常可以基于距离和字符来构建进化树。距离法分为非加权分组平均法（UPGMA）和邻接法（NJ）；字符法分为最大简约法（MP）、最大似然法（ML）和贝叶斯马尔科夫链蒙特卡洛法（BMCMC）。

2. 参考答案

系统发育树可以分为 Newick、NEXUS、NHX、Jplace 等格式。

Newick 格式因为其被许多软件支持，所以是最广泛使用的进化树的格式。

Newick 树文本格式详述：Newick 树文本以分号结尾，其内部节点由一对匹配的括号表示，而括号内是该节点的后代节点，如（t2:0.04,t1:0.34）表示为 t2 和 t1 的父节点，t2、t1 均为其直系后代。同级的内部节点用逗号（,）分隔，同时叶节点用它们的名称表示。枝长（从父节点到子节点）由子节点后面的实数表示，两者之间用冒号（:）分隔。与内部节点或分支相关联的单一数据（如自举值）可以被编码为节点标签，并用冒号（:）前面简单的文本或数字表示。

3. 参考答案

（1）read.tree()。

（2）read.raxml()。

（3）read.beast()。

（4）read.mrbayes()。

（5）NEXUS：read.hyphy()。

祖先序列：read.hyphy.seq()。

（6）mlc:read.codeml_mlc()。

rst:read.paml_rst() 或 read.codeml()。

（7）NEXUS:read.mega()。

read.mega_tubular()。

（8）read.r8S()。

第 2 章

1. 参考答案

将 phylo 对象转换为 tibble：as.tibble()，将 phylo 对象转换为 treedata：as.treedata()。
将 treedata 对象转换为 tibble：as.tibble()，将 treedata 对象转换为 phylo：as.phylo()。

2. 参考答案

```
library(treeio)
set.seed(123)
rtree <- rtree(23)
## 随机选定一个叶子节点
tn <- sample(1:Ntip(rtree), 1)
ancestor(rtree,tn)
## 随机选定一个内部节点
nn <- sample((Ntip(rtree)+1):(Ntip(rtree)+Nnode(rtree)), 1)
offspring(rtree,nn)
```

3. 参考答案

```
library(ggtree)
library(treeio)
set.seed(123)
rtree <- rtree(23)
node <- sample((Ntip(rtree)+1):(Ntip(rtree)+Nnode(rtree)),1)
tr1 <- tree_subset(rtree, node, levels_back = 0)
ggtree(tr1) + geom_tiplab()
tip <- sample(rtree$tip.label,1)
tr2 <- tree_subset(rtree,tip,levels_back = 1)
ggtree(tr2) + geom_tiplab()
```

第 3 章

1. 参考答案

```
library(treeio)
t1 <- read.beast("beast-file-path")
t2 <- read.NHX("NHX-file-path")
mt <- merge_tree(t1, t2)
write.beast(mt)
write.tree(mt)
```

2. 参考答案

```
read.jtree()     ## 导入 jtree 对象
write.jtree()    ## 导出 jtree 对象
```

第 4 章

1. 参考答案

```
library(ggtree)
set.seed(123)
tr <- rtree(15)
p <- ggtree(tr, layout = "roundrect") + geom_tiplab(angle = -45)
ggplotify::as.ggplot(p, angle = 45)
```

2. 参考答案

```
library(ggtree)
library(phytools)
library(ggplot2)
set.seed(123)
tr <- rtree(30)
t <- fastBM(tr, nsim = 1)
df <- data.frame(node = fortify(tr)$node ,traits = c(t, fastAnc(tr,t)))
tree <- full_join(tr, df, by = "node")
ggtree(tree, size = 3) +
    geom_tree(aes(color = traits), continuous = "color", size = 2) +
    scale_color_gradientn(colours=c("red", 'orange', 'green', 'cyan', 'blue'))
```

3. 参考答案

```
library(ggtree)
set.seed(123)
tr <- rtree(20)
p1 <- ggtree(tr, layout = "circular", color = "transparent", branch.length = "none")
p1 + geom_nodepoint(color = "red", alpha = 0.5) +
    geom_tippoint(color = "yellow", shape = 8) +
    theme_tree("black")
```

第 5 章

1. 参考答案

使用图形语法对进化树进行可视化能够轻松地生成一些简单的图层，做到快速对进化树进行基础的可视化。同时，通过组合这些简单的图层，我们还能够生成更为复杂的图形，以做到对进化树进行充分的注释。

2. 参考答案

```
library(ggtree)
set.seed(123)
tr <- rtree(20)
dat <- data.frame(f = c(7, 2, 1),
            t = c(11, 15, 9),
            color = c("red", "blue", "darkgreen"),
            size = c(3, 2, 1)
            )
ggtree(tr, layout = "inward_circular", xlim=c(5, 0)) +
  geom_highlight(node = 23, fill = "steelblue") +
  geom_taxalink(data = dat,
            mapping = aes(taxa1 = f,
                    taxa2 = t,
                    color = I(color),
                    size = I(size)))
```

3. 参考答案

（1）使用 geom_range 以红色水平条带的形式来将（如利用 BEAST 推断的枝长的）置信区间注释于对应的节点上。

（2）使用 geom_text() 图层函数或 geom_labelt() 图层函数将（如利用 BEAST 推

断的后验概率或利用 HyPhy、CODEML 等软件得出的氨基酸替换等）文本信息注释于分支上。

（3）通过对树着色的方式来呈现连续型（如利用 CODEML 推断的 dN/dS）和离散型（如物种饮食偏好、祖先状态可能等预测推断）数据。

第 6 章

1. 参考答案

```
library(ggtree)
library(tidytree)
set.seed(123)
tr <- rtree(20)
## 找到子节点
## 示例 29, 30
child(tr, 28)
tr2 <- groupClade(tr, c(29,30))
p <- ggtree(tr2, aes(color=group)) +
    theme(legend.position='none') +
    scale_color_manual(values=c("black", "firebrick", "steelblue"))
p1 <- scaleClade(p, node = 30, scale = 3)
p2 <- collapse(p1, node = 29) +
    geom_point2(aes(subset=(node==29)), shape=21, size=3, fill='red')
```

2. 参考答案

```
library(ggtree)
set.seed(123)
tr <- rtree(20)
p <- ggtree(tr)
p1 <- rotate(p, 24)
p2 <- open_tree(p1, 90)
p3 <- rotate_tree(p2, 270)
```

第 7 章

1. 参考答案

（1）示例如下。

```
library(ggtree)
set.seed(123)
```

```r
x=rtree(4)
p=ggtree(x)+geom_tippoint()
# 外部数据
data <- data.frame(label=x$tip.label,fake_trait=1:Ntip(x),fake_trait2=c(1,1,2,2))
data
# 与外部数据合并
p1=p %<+% data
# 映射外部数据到树上
p1+geom_tippoint(aes(size=fake_trait,color=fake_trait2))
```

（2）示例如下。

```r
# 利用geom_facet()图层函数生成组合图
p+geom_facet(panel="fake_data",data=data,geom= geom_point,aes(x=fake_trait,color=fake_trait2,fill=fake_trait2))
# 利用facet_plot()函数生成组合图
facet_plot(p,panel = "fake_data",data=data,geom = geom_point,aes(x=fake_trait,color=fake_trait2,fill=fake_trait2))
```

（3）示例如下。

```r
# 生成热图数据
htdt <- data.frame(label=x$tip.label,s1=rnorm(Ntip(x)),s2=rnorm(Ntip(x)),s3=rnorm(Ntip(x)))
htdt2 <- htdt[,-1]
rownames(htdt2) <- htdt[,1]
# 可视化热图数据
gheatmap(p,htdt2,width=.5,offset =.2)
```

第8章

1. 参考答案

```r
library(ggtree)
set.seed(123)
tr <- rtree(5)
path <- data.frame(node = 1:5,
                   path = c("path_to_node1_pic",
                            "path_to_node2_pic ",
                            "path_to_node3_pic ",
                            "path_to_node4_pic ",
                            "path_to_node5_pic "))
```

```
tri <- full_join(tr, path, by = "node")

ggtree(tri) +
    geom_tiplab(aes(image = path), geom="image")
```

2. 参考答案

```
library(ggtree)
library(ggplot2)
data("mtcars")
set.seed(123)
tr <- rtree(5)
p <- ggtree(tr)
p1 <- ggplot(mtcars,aes(factor(cyl),fill=factor(am))) +
        geom_bar() +
        theme_void() + theme(legend.position='none')
p2 <- ggplot(mtcars,aes(mpg,fill=factor(vs))) +
        geom_density() +
        theme_void() + theme(legend.position='none')
p3 <- ggplot(mtcars,aes(factor(gear),fill=factor(cyl))) +
        geom_bar(position="dodge") +
        theme_void() + theme(legend.position='none')
plist <- list("8"= p1, "9"=p2, "7" = p3)
pi <- p + geom_inset(plist)
```

3. 无

第 9 章

1. 参考答案

ggtree 可以衔接 phylobase 包输出的 phylo4 对象及 phylo4d 对象，ade4 包输出的是 phylog 对象，phyloseq 包输出的是 phyloseq 对象。

示例：构建 phylo4d 对象并将其进行可视化。

```
library(ggtree)
library(phylobase)
set.seed(123)
x=rtree(4)
data<- data.frame(label=x$tip.label,
                  s1=seq(4.5,5,length.out=4),
```

```
                     s2=seq(5,5.5,length.out=4),
                     s3=seq(5.5,6,length.out=4))
d2 <- data[,-1]
rownames(d2) <- data[,1]
phy1 <- as(x, "phylo4")
phyd <- phylo4d(phy1, d2)
ggtree(phyd) +
  geom_tiplab(align = T) +
  geom_tippoint(aes(color = s1,size = s1),x = 2.2,shape = 1) +
  geom_tippoint(aes(color = s2,size = s2),x = 2.4,shape = 1) +
  geom_tippoint(aes(color = s3,size = s3),x = 2.6,shape = 1) +
  xlim(0, 2.8)
```

2. 参考答案

```
library(ggtree)
hc <- hclust(dist(UScitiesD))
  den <- as.dendrogram(hc)

ggtree(den, layout = 'dendrogram') +
       geom_tiplab() +
       theme_dendrogram()
```

第 10 章

1. 参考答案

```
library(ggtreeExtra)
# p 为 ggtree 对象
p <- ggtree(tree)
p + geom_fruit(
    data = long_format_data,
    geom = geom_xx,
    mapping = aes(x = value, y = tip.id, other_aes),
    pwidth = .2
 )
```

2. 参考答案

```
library(TDbook)
data(tree_hmptree)
data(df_barplot_attr)
```

```
library(TDbook)
library(ggplot2)
library(ggtree)
library(ggtreeExtra)
p <- ggtree(tree_hmptree)
p <- p +
    geom_fruit(
    data = df_barplot_attr,
    geom = geom_col,
    mapping = aes(x = HigherAbundance, y = ID, fill = Sites),
    pwidth = .3 ,
    axis.params=list(axis='x'),
    grid.params=list())
```

3. 参考答案

用户可以使用 ggnewscale 包中的 new_scale("fill") 指令来添加新的属性映射。除了 fill，还可以相应地使用诸如 size、color、alpha 等其他的美学映射属性来替代，添加相应的属性映射。

4. 参考答案

总共有 3 种方式：第一种，整理成对应 geom_xx() 图层函数所需的 long format 数据框，通过 geom_fruit() 图层函数中的 data 导入；第二种，通过 left_join() 函数把 long format 数据框添加到 treedata 或 phylo 中（有时候需要对数据进行展开操作可以把 td_unnest 传送到 data 参数中）。第三种，对于没有重复 id（tip 或者 node）的数据框可以先通过 %<+% 操作符将数据整合到 ggtree 对象内，再进行可视化。

第 11 章

参考答案

```
rp <- BiocManager::repositories()
db <- utils::available.packages(repo=rp)
x <- tools::package_dependencies('ggtree', db=db,
                                which = c("Depends", "Imports"),
                                reverse=TRUE)
print(x)
```

第 12 章

1. 参考答案

facet_widths() 函数可以用于调节分面面板的尺寸。

facet_labeller() 函数可以用于调节分面面板的标签。

当两者连用时，要先使用 facet_labeller() 函数调节标签，再使用 facet_widths() 函数调节尺寸。

2. 参考答案

```
library(ggtree)
library(ggplot2)
ggplot(iris, aes(x = Sepal.Width, y = Petal.Length)) +
  geom_point2(aes(subset = Petal.Length>6 | Sepal.Width<3))
```

3. 参考答案

```
ggtree(x) + layout_circular()
ggtree(x) + layout_inward_circular()
ggtree(x) + layout_fan()
```

4. 参考答案

```
library(ggtree)
library(ggplot2)
data(mpg)
p <- ggplot(data = mpg, mapping = aes(x = displ, y = hwy)) +
     geom_point() +
     facet_grid(.~class,scales = "free_x")
p + xlim_expand(c(0, 10), "2seater")
```

5. 参考答案

ggexpand 可以单独缩放 x 轴或 y 轴，也可以同时缩放 x 轴与 y 轴。

ggexpand() 函数用于接收以下 3 个参数。

- ratiao：按 x 轴或 y 轴范围的比例缩放图片坐标。
- derection：其值为 1 或 -1。当值为 1 时表示向右侧延伸 x 轴，当值为 -1 时表示向左侧延伸 x 轴。
- side：其值为 h、v 或 hv。h 表示水平横向延申，v 表示垂直纵向延申，hv 表示同时进行水平和垂直方向延伸。

6. 参考答案

```
library(ggtree)
library(ggplot2)
rtree <- rtree(23)
p <- ggtree(rtree)
nn <- sample((Ntip(rtree)+1):(Ntip(rtree)+Nnode(rtree)), 1)
get_taxa_name(p, node = nn)
```

7. 参考答案

在默认情况下，使用 label_pad() 函数会生成"."占位，用于保证叶节点标签末端对齐。

反侵权盗版声明

电子工业出版社依法对本作品享有专有出版权。任何未经权利人书面许可，复制、销售或通过信息网络传播本作品的行为；歪曲、篡改、剽窃本作品的行为，均违反《中华人民共和国著作权法》，其行为人应承担相应的民事责任和行政责任，构成犯罪的，将被依法追究刑事责任。

为了维护市场秩序，保护权利人的合法权益，我社将依法查处和打击侵权盗版的单位和个人。欢迎社会各界人士积极举报侵权盗版行为，本社将奖励举报有功人员，并保证举报人的信息不被泄露。

举报电话：（010）88254396；（010）88258888
传　　真：（010）88254397
E-mail：dbqq@phei.com.cn
通信地址：北京市万寿路173信箱
　　　　　电子工业出版社总编办公室
邮　　编：100036